Introduction to Agricultural Chemistry

Introduction to
Agricultural Chemistry

Trinity Gill

WILLFORD PRESS

www.willfordpress.com

Published by Willford Press,
118-35 Queens Blvd., Suite 400,
Forest Hills, NY 11375, USA

ISBN: 978-1-68285-999-5

Cataloging-in-Publication Data

Introduction to agricultural chemistry / Trinity Gill.
    p. cm.
Includes bibliographical references and index.
ISBN 978-1-68285-999-5
1. Agricultural chemistry. 2. Chemistry. I. Gill, Trinity.
S585 .I58 2022
630.24--dc23

For information on all Willford Press publications
visit our website at www.willfordpress.com

WILLFORD PRESS

Table of Contents

Preface

This book is a culmination of my many years of practice in this field. I attribute the success of this book to my support group. I would like to thank my parents who have showered me with unconditional love and support and my peers and professors for their constant guidance.

Agricultural chemistry is involved in the study of chemistry and biochemistry in relation to agricultural production. It also focuses on the processing of raw products into foods and beverages as well as in environmental monitoring and remediation. It observes the relationships between plants, animals and bacteria with their environment. Agricultural chemistry studies the varied processes by which humans obtain food and fibers. It also works to control these processes in order to increase yields, improve quality and reduce costs. This textbook aims to shed light on some of the unexplored aspects of agricultural chemistry. While understanding the long-term perspectives of the topics, the book makes an effort in highlighting their impact as a modern tool for the growth of the discipline. It will serve as a valuable source of reference for those interested in this field.

The details of chapters are provided below for a progressive learning:

Chapter – What is Agricultural Chemistry?

The branch of chemistry which focuses on agricultural production, processing of raw foods into food and environmental monitoring is termed as agricultural chemistry. Some of the different kinds of soil studied within this discipline are sandy soil, silt soil, clay soil and loamy soil. This chapter has been carefully written to provide an easy introduction to the varied facets of agricultural chemistry.

Chapter – Chemistry of Soil

The study of the chemical characteristics of soil is termed as soil chemistry. Some of the major areas of study within this field are soil pH, soil reaction, soil acidity and soil alkalinity. The topics elaborated in this chapter will help in gaining a better perspective about these focus areas of soil chemistry.

Chapter – Soil Composition and Ionic Reactions

Soil has numerous chemical compounds present in it such as phosphoric acid, calcium and acid sulfate. One of the most important ionic reaction which takes place in the soil is the exchange reaction. This chapter closely examines these key components as well as ionic reactions to provide an extensive understanding of the subject.

Chapter – Manures and Fertilizers

The organic matter which is generally derived from animal feces is termed as manure. Any material of synthetic or natural origin which is applied to soil or plant tissues to aid the growth of plants is termed as a fertilizer. All the diverse principles of manures and fertilizers as well as their different types have been carefully analyzed in this chapter.

Chapter – Common Agrochemicals

The chemical compounds which are used in agriculture are termed as agrochemicals. Some of the commonly used agrochemicals are insecticides, pesticides, biopesticides, herbicides, bioherbicides and fungicides. The diverse aspects of these agrochemicals have been thoroughly discussed in this chapter.

Trinity Gill

1

What is Agricultural Chemistry?

The branch of chemistry which focuses on agricultural production, processing of raw foods into food and environmental monitoring is termed as agricultural chemistry. Some of the different kinds of soil studied within this discipline are sandy soil, silt soil, clay soil and loamy soil. This chapter has been carefully written to provide an easy introduction to the varied facets of agricultural chemistry.

Agricultural Chemistry is the study of chemistry and biochemistry in relation to the agricultural field. In agricultural chemistry, factors such as agricultural production, the utilization of agricultural products, and environmental matters are studied and ways to improve them are developed. In agricultural chemistry, the relationship between plants animals and the environment is emphasized to procure improvements in the agricultural sector.

With their knowledge of biochemistry, a biochemist employs techniques can be used to improve the production, protection, and use of livestock and crops. As a form of applied sciences, some of the main aims of agricultural chemistry are:

- Increase yield of crops and livestock.

- Improving the quality of products.

- Reducing cost of products.

Uses

To achieve the above-mentioned goals, agricultural chemistry takes up a lot of techniques. Some of them are:

- Improving the quality of soil: Agricultural chemists work on preserving the quality of soil and also improving the quality of the soil.

- Developing effective materials such as fungicides, insecticides, herbicides, and other pesticides. Also certain chemical materials such as animal feed supplements, fertilizers, and plant growth regulators.

- Sustainability of our needs: Agricultural chemists also look to sustain our resources so as to ensure we don't use up all of our resources but replenish it for the future generations as well.

Soil

Soil is the biologically active, porous medium that has developed in the uppermost layer of Earth's crust. Soil is one of the principal substrata of life on Earth, serving as a reservoir of water and nutrients, as a medium for the filtration and breakdown of injurious wastes, and as a participant in the cycling of carbon and other elements through the global ecosystem. It has evolved through weathering processes driven by biological, climatic, geologic, and topographic influences.

Since the rise of agriculture and forestry in the 8th millennium BCE, there has also arisen by necessity a practical awareness of soils and their management. In the 18th and 19th centuries the Industrial Revolution brought increasing pressure on soil to produce raw materials demanded by commerce, while the development of quantitative science offered new opportunities for improved soil management. The study of soil as a separate scientific discipline began about the same time with systematic investigations of substances that enhance plant growth. This initial inquiry has expanded to an understanding of soils as complex, dynamic, biogeochemical systems that are vital to the life cycles of terrestrial vegetation and soil-inhabiting organisms—and by extension to the human race as well.

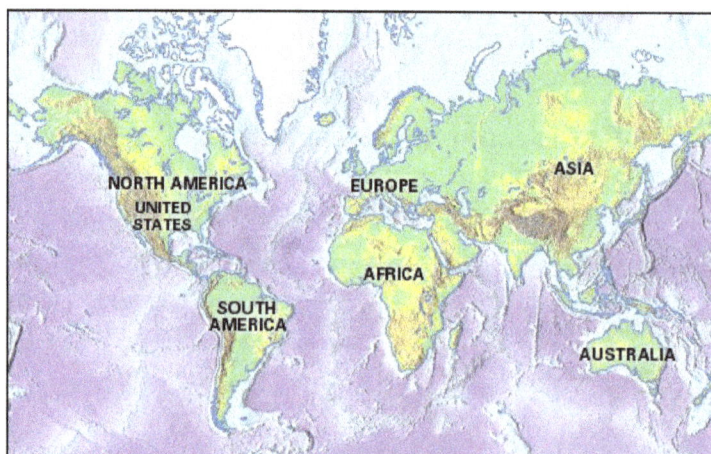

Soil map of the world.

Soil Profile

Soil Horizons

Soils differ widely in their properties because of geologic and climatic variation over distance and time. Even a simple property, such as the soil thickness, can range from a few centimetres to many metres, depending on the intensity and duration of weathering, episodes of soil deposition and erosion, and the patterns of landscape evolution. Nevertheless, in spite of this variability, soils have a unique structural characteristic that distinguishes them from mere earth materials and serves as a basis for their classification: a vertical sequence of layers produced by the combined actions of percolating waters and living organisms.

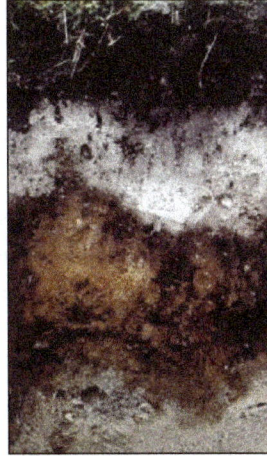

Podzol soil profile from Ireland, showing a bleached layer from which
humus and metal oxides have been leached and subsequently deposited in the typically
reddish horizon below.

These layers are called horizons, and the full vertical sequence of horizons constitutes the soil profile. Soil horizons are defined by features that reflect soil-forming processes. For instance, the uppermost soil layer (not including surface litter) is termed the A horizon. This is a weathered layer that contains an accumulation of humus (decomposed, dark-coloured, carbon-rich matter) and microbial biomass that is mixed with small-grained minerals to form aggregate structures.

In the figure below, below A lies the B horizon. In mature soils this layer is characterized by an accumulation of clay (small particles less than 0.002 mm [0.00008 inch] in diameter) that has either been deposited out of percolating waters or precipitated by chemical processes involving dissolved products of weathering. Clay endows B horizons with an array of diverse structural features (blocks, columns, and prisms) formed from small clay particles that can be linked together in various configurations as the horizon evolves.

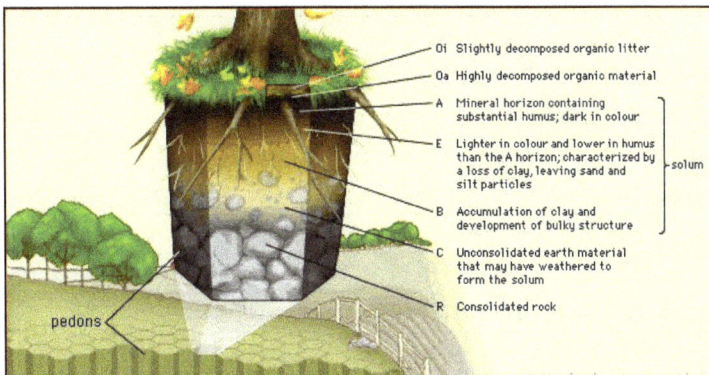

The soil profile, showing the major layers from the O horizon (organic material) to the R horizon (consolidated rock). A pedon is the smallest unit of land surface that can be used to study the characteristic soil profile of a landscape.

Below the A and B horizons is the C horizon, a zone of little or no humus accumulation or soil structure development. The C horizon often is composed of unconsolidated parent material from which the A and B horizons have formed. It lacks the characteristic features of the A and B horizons and may be either relatively unweathered or deeply weathered. At some depth below the A, B, and C horizons lies consolidated rock, which makes up the R horizon.

These simple letter designations are supplemented in two ways. First, two additional horizons are defined. Litter and decomposed organic matter (for example, plant and animal remains) that typically lie exposed on the land surface above the A horizon are given the designation O horizon, whereas the layer immediately below an A horizon that has been extensively leached (that is, slowly washed of certain contents by the action of percolating water) is given the separate designation E horizon, or zone of eluviation. The development of E horizons is favoured by high rainfall and sandy parent material, two factors that help to ensure extensive water percolation. The solid particles lost through leaching are deposited in the B horizon, which then can be regarded as a zone of illuviation.

Soil horizon letter designations	
Base symbols for surface horizons	
O	Organic horizon containing litter and decomposed organic matter
A	Mineral horizon darkened by humus accumulation
Base symbols for subsurface horizons	
E	Mineral horizon lighter in colour than an A or O horizon and depleted in clay minerals
AB or EB	Transitional horizon more like A or E than B
BA or BE	Transitional horizon more like B than A or E
B	Accumulated clay and humus below the A or E horizon
BC or CB	Transitional horizon from B to C
C	Unconsolidated earth material below the A or B horizon
R	Consolidated rock
Suffixes added for special features of horizons	
a	Highly decomposed organic matter
b	Buried horizon
c	Concretions or hard nodules (iron, aluminum, manganese, or titanium)
e	Organic matter of intermediate decomposition
f	Frozen soil
g	Gray colour with strong mottling and poor drainage
h	Accumulation of organic matter
i	Slightly decomposed organic matter
k	Accumulation of carbonate
m	Cementation or induration
n	Accumulation of sodium
o	Accumulation of oxides of iron and aluminum
p	Plowing or other anthropogenic disturbance

q	Accumulation of silica
r	Weathered or soft bedrock
s	Accumulation of metal oxides and organic matter
t	Accumulation of clay
v	Plinthite (hard iron-enriched subsoil material)
w	Development of colour or structure
x	Fragipan character (high-density, brittle)
y	Accumulation of gypsum
z	Accumulation of salts

The combined A, E, B horizon sequence is called the solum. The solum is the true seat of soil-forming processes and is the principal habitat for soil organisms. (Transitional layers, having intermediate properties, are designated with the two letters of the adjacent horizons.)

The second enhancement to soil horizon nomenclature is the use of lowercase suffixes to designate special features that are important to soil development. The most common of these suffixes are applied to B horizons: g to denote mottling caused by waterlogging, h to denote the illuvial accumulation of humus, k to denote carbonate mineral precipitates, o to denote residual metal oxides, s to denote the illuvial accumulation of metal oxides and humus, and t to denote the accumulation of clay.

Pedons and Polypedons

Soils are natural elements of weathered landscapes whose properties may vary spatially. For scientific study, however, it is useful to think of soils as unions of modules known as pedons. A pedon is the smallest element of landscape that can be called soil. Its depth limit is the somewhat arbitrary boundary between soil and "not soil" (e.g., bedrock). Its lateral dimensions must be large enough to permit a study of any horizons present—in general, an area from 1 to 10 square metres (10 to 100 square feet), taking into account that a horizon may be variable in thickness or even discontinuous. Wherever horizons are cyclic and recur at intervals of 2 to 7 metres (7 to 23 feet), the pedon includes one-half the cycle. Thus, each pedon includes the range of horizon variability that occurs within small areas. Wherever the cycle is less than 2 metres, or wherever all horizons are continuous and of uniform thickness, the pedon has an area of 1 square metre.

Soils are encountered on the landscape as groups of similar pedons, called polypedons, that contain sufficient area to qualify as a taxonomic unit. Polypedons are bounded from below by "not soil" and laterally by pedons of dissimilar characteristics.

Soil Behaviour

Grain Size and Porosity

The grain size of soil particles and the aggregate structures they form affect the ability of a soil to transport and retain water, air, and nutrients. Grain size is classified as clay if the particle diameter is less than 0.002 mm (0.0008 inch), as silt if it is between 0.002 mm (0.0008 inch) and 0.05 mm (0.002 inch), or as sand if it is between 0.05 mm (0.002 inch) and 2 mm (0.08 inch). Soil texture refers to the relative proportions of sand, silt, and clay particle sizes, irrespective of chemical or mineralogical composition. Sandy soils are called coarse-textured, and clay-rich soils are called fine-textured. Loam is a textural class representing about one-fifth clay, with sand and silt sharing the remainder equally.

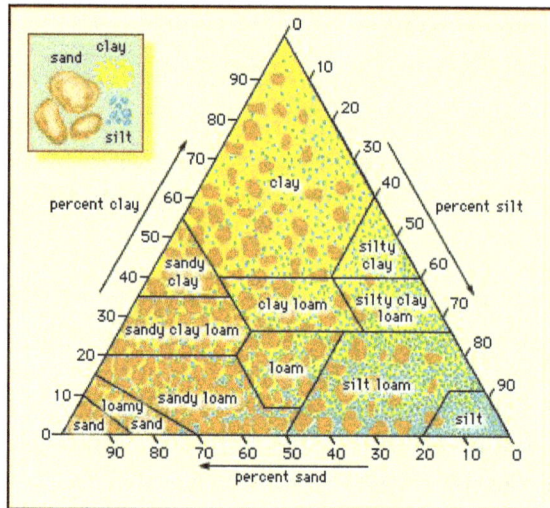

Soil texture as a function of the proportion of sand, silt, and clay particle sizes.

Pore radii (space between soil particles) can range from millimetre-scale between sand grains to micrometre-scale between clay grains. Soil particles falling into the three principal size categories may have various mineralogical or chemical compositions, although sand particles often are composed of quartz and feldspars, silt particles often are micaceous, and clay particles often contain layer-type aluminosilicates (the so-called clay minerals). Organic matter and amorphous mineral matter also are important constituents of soil clay particles.

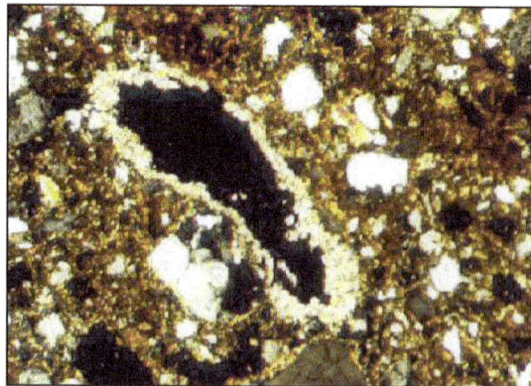

Microscopic view of an Inceptisol, showing small crystallites of carbonate minerals (around the central black void), quartz sand grains (white), and iron oxides and organic matter (dark brown).

Porosity reflects the capacity of soil to hold air and water, and permeability describes the ease of transport of fluids and their dissolved components. The porosity of a soil horizon increases as its texture becomes finer, whereas the permeability decreases as the average pore size becomes smaller. Small pores not only restrict the passage of matter, but they also bring it into close proximity with chemical binding sites on the particle surface that can slow its movement. Clay and humus affect both soil porosity and permeability by binding soil grains together into aggregates, thereby creating a network of larger pores (macropores) that facilitate the movement of water. Plant roots open pores between soil aggregates, and cycles of wetting and drying create channels that allow water to pass easily. (However, this structure collapses under waterlogging conditions.) The stability of aggregates increases with humus content, especially humus that originates from grass vegetation. For soils that are not disturbed significantly by human activities, however, the pore space and the varieties of macropores are more important determinants of porosity than the soil texture. As a general rule, average pore size decreases from certain agricultural practices and other human uses of soil.

Water Runoff

Aggregates of soil particles whose formation has not been influenced by human intervention are called peds. The peds in the surface horizons of soils develop into clods under the effects of cultivation and the traffic of urbanization. Soils whose A horizon is dense and unstructured increase the fraction of precipitation that will become surface runoff and have a high potential for erosion and flooding. These soils include not only those whose peds have been degraded but also coarse-textured soils with low porosity, particularly those of arid regions.

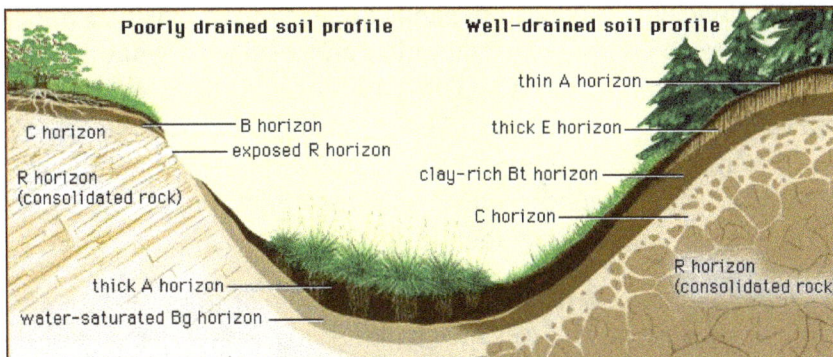

Soil profiles on hillslopes.

The thickness and composition of soil horizons vary with position on a hillslope and with water drainage. For example, on the upper slopes of poorly drained profiles, underlying rock may be exposed by surface erosion, and nutrient-rich soils (A horizon) may accumulate at the toeslope. On the other hand, in well-drained profiles under forest cover, the leached layers (E horizon) may be relatively thick and surface erosion minimal.

A well-developed clay horizon (Bt) presents a deep-lying obstacle to the downward perco-lation of water. Subsurface runoff cannot easily penetrate the clay layer and flows laterally along the horizon as it moves toward the stream system. This type of runoff is slower than its erosive counterpart over the land surface and leads to water saturation of the upper part of the soil profile and the possibility of gravity-induced mass movement on hillslopes (e.g., landslides). It is also responsible for the translocation (migration) of dissolved products of chemical weathering down a hillslope sequence of related soil profiles (a toposequence). Subsurface water flow is also influenced by macropores, which, as noted above, are creat-ed through plant root growth and decay, animal burrowing activities, soil shrinkage while drying, or fracturing. In general, subsurface runoff processes are characteristic of soils in humid regions, whereas surface runoff is characteristic of arid regions and, of course, any landscape altered significantly by cultivation or urbanization.

Chemical Characteristics

Mineral Content

The bulk of soil consists of mineral particles that are composed of arrays of silicate ions (SiO_4^{4-}) combined with various positively charged metal ions. It is the number and type of the metal ions present that determine the particular mineral. The most common mineral found in Earth's crust is feldspar, an aluminosilicate that contains sodium, potassium, or calcium (sometimes called bases) in addition to aluminum ions. Weath-ering breaks up crystals of feldspars and other silicate minerals and releases chemical compounds such as bases, silica, and oxides of iron and aluminum (Fe_2O_3 and alumina [Al_2O_3]). After the bases are removed by leaching, the remaining silica and alumina combine to form crystalline clays.

The kind of crystalline clay produced depends on leaching intensity. Prolonged leach-ing leaves little silica to combine with alumina and results in what are known as 1:1 clays, consisting of alternating silica and alumina sheets; less extensive leaching leads to the formation of 2:1 clays, consisting of one alumina sheet sandwiched between two silica sheets. In neither case is the result solely one of the two types, though 1:1 clay is predominant in the tropics after prolonged leaching and 2:1 clay more abundant when leaching is less extensive in more temperate climates.

The solid soil particles are chemically reactive because of the presence of electrically charged sites on their surfaces. If a reactive site binds a dissolved ion or molecule to form a stable unit, a "surface complex" is said to exist. The formation reaction itself is called surface complexation. Surface complexation is an example of adsorption, a chemical pro-cess in which matter accumulates on a solid particle surface. Ions such as Ca^{2+} (calcium), Mg^{2+} (magnesium), Na^+(sodium), and NO_3^- (nitrate) do not tend to adsorb strongly, mak-ing these important plant nutrients susceptible to easy replacement. Once ejected from their surface sites, these ions may be leached downward by percolating water to become removed from the biogeochemical cycles occurring in the upper part of the soil profile.

Ferralsol soil profile from Brazil, showing a deep red subsurface horizon
resulting from accumulations of iron and aluminum oxides.

Freshwater leaching of soils brings hydrogen ions (H^+) that increase mineral solubility, releasing Al^{3+} (aluminum), a toxic ion that can displace nutrients such as Ca^{2+}. The gradual loss of nutrients and the accumulation of adsorbed H^+ and Al^{3+} characterize the build-up of soil acidity, with its harmful effects on organisms. Soils display their acidity by a decrease in content of acid-soluble minerals (for example, feldspars or clay minerals) and an increase in insoluble minerals (iron and aluminum oxides). Soils weathered by freshwater leaching evolve from clay particles with a prevalence of metal ion-binding sites to highly weathered metal oxides that do not have sites that bind readily with metal ions.

Organic Content

Mollisol soil profile, showing a typically dark surface horizon rich in humus.

The second major component of soils is organic matter produced by organisms. The total organic matter in soil, except for materials identifiable as undecomposed or partially decomposed biomass, is called humus. This solid, dark-coloured component of soil plays a significant role in the control of soil acidity, in the cycling of nutrients, and in the detoxification of hazardous compounds. Humus consists of biological molecules such as proteins and carbohydrates as well as the humic substances (polymeric compounds produced through microbial action that differ from metabolically active compounds).

The processes by which humus forms are not fully understood, but there is agreement that four stages of development occur in the transformation of soil biomass to humus: decomposition of biomass into simple organic compounds, metabolization of the simple compounds by microbes, cycling of carbon, hydrogen, nitrogen, and oxygen between soil organic matter and the microbial biomass, and microbe-mediated polymerization of the cycled organic compounds.

The investigation of molecular structure in humic substances is a difficult area of current research. Although it is not possible to describe the molecular configuration of humic substances in any but the most general terms, these molecules contain hydrogen ions that dissociate in fresh water to form molecules that bear a net negative charge. These negatively charged sites can interact with toxic metal ions and effectively remove them from further interaction with the environment.

Much of the molecular framework of soil organic matter, however, is not electrically charged. The uncharged portions of humic substances can react with synthetic organic compounds such as pesticides, fertilizers, solid and liquid waste materials, and their degradation products. Humus, either as a separate solid phase or as a coating on mineral surfaces, can immobilize these compounds and, in some instances, detoxify them significantly.

Biological Phenomena

Fertile soils are biological environments teeming with life on all size scales, from microfauna (with body widths less than 0.1 mm [0.004 inch]) to mesofauna (up to 2 mm [0.08 inch] wide) and macrofauna (up to 20 mm [0.8 inch] wide). The most numerous soil organisms are the unicellular microfauna: 1 kilogram (2.2 pounds) of soil may contain 500 billion bacteria, 10 billion actinomycetes (filamentous bacteria, some of which produce antibiotics), and nearly 1 billion fungi. The multicellular animal population can approach 500 million in a kilogram of soil, with microscopic nematodes (roundworms) the most abundant. Mites and springtails, which are categorized as mesofauna, are the next most prevalent. Earthworms, millipedes, centipedes, and insects make up most of the rest of the larger soil animal species. Plant roots also make a significant contribution to the biomass—the combined root length from a single plant can exceed 600 km (373 miles) in the top metre of a soil profile.

The soil flora and fauna play an important role in soil development. Microbiological activity in the rooting zone of soils is important to soil acidity and to the cycling of nutrients. Aerobic and anaerobic (oxygen-depleted) microniches support microbes that determine the rate of the production of carbon dioxide (CO_2) from organic matter or of nitrate (NO_3^-) from molecular nitrogen (N_2).

The carbon and nitrogen cycles are two important microbe-mediated cycles, it is worth pointing out how they illustrate the complex, integrated nature of a soil's physical, chemical, and biological behaviour: soil peds and pore spaces provide microniches for the action of carbon- and nitrogen-cycling organisms, soil humus provides the nutrient reservoirs, and soil biomassprovides the chemical pathways for cycling. The carbon in dead biomass is converted to CO_2 by aerobic microorganisms and to organic acids or alcohols by anaerobic microorganisms. Under highly anaerobic conditions, methane (CH_4) is produced by bacteria. The CO_2 produced can be used by photosynthetic microorganisms or by higher plants to create new biomass and thus initiate the carbon cycle again.

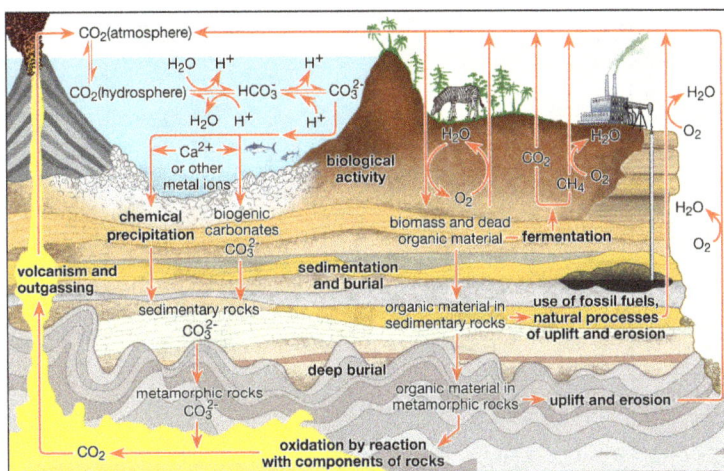

The carbon cycle.

Carbon is transported in various forms through the atmosphere, the hydrosphere, and geologic formations. One of the primary pathways for the exchange of carbon dioxide (CO_2) takes place between the atmosphere and the oceans; there a fraction of the CO_2 combines with water, forming carbonic acid (H_2CO_3) that subsequently loses hydrogen ions (H^+) to form bicarbonate (HCO_3^-) and carbonate (CO_3^{2-}) ions. Mollusk shells or mineral precipitates that form by the reaction of calcium or other metal ions with carbonate may become buried in geologic strata and eventually release CO_2 through volcanic outgassing. Carbon dioxide also exchanges through photosynthesis in plants and through respiration in animals. Dead and decaying organic matter may ferment and release CO_2 or methane (CH_4) or may be incorporated into sedimentary rock, where it is converted to fossil fuels. Burning of hydrocarbon fuels returns CO_2 and water (H_2O) to the atmosphere. The biological and anthropogenic pathways are much faster than the geochemical pathways and, consequently, have a greater impact on the composition and temperature of the atmosphere.

The nitrogen (N) bound into proteins in dead biomass is consumed by microorganisms and converted into ammonium ions (NH_4^+) that can be directly absorbed by plant roots (for example, lowland rice). The ammonium ions are usually converted to nitrite ions

(NO_2^-) by Nitrosomonas bacteria, followed by a second conversion to nitrate (NO_3^-) by Nitrobacterbacteria. This very mobile form of nitrogen is that most commonly absorbed by plant roots, as well as by microorganisms in soil. To close the nitrogen cycle, nitrogen gas in the atmosphere is converted to biomass nitrogen by Rhizobium bacteria living in the root tissues of legumes (e.g., alfalfa, peas, and beans) and leguminous trees (such as alder) and by cyanobacteria and Azotobacter bacteria.

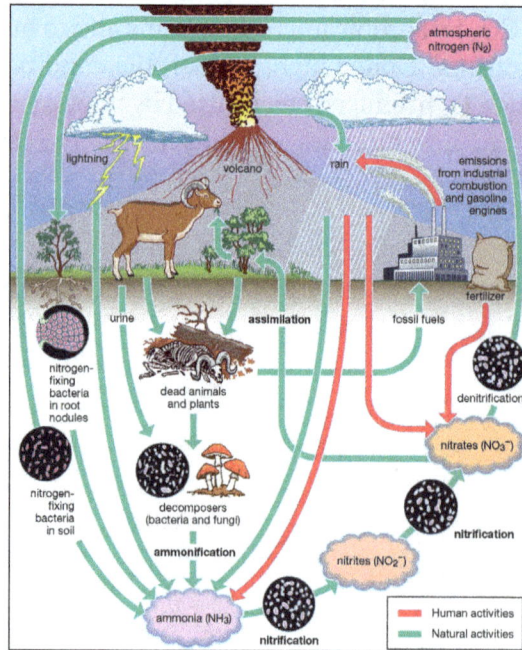

The nitrogen cycle.

Soil Formation

Soils evolve under the action of biological, climatic, geologic, and topographic influences. The evolution of soils and their properties is called soil formation, and pedologists have identified five fundamental soil formation processes that influence soil properties. These five "state factors" are parent material, topography, climate, organisms, and time.

Parent Material

Parent material is the initial state of the solid matter making up a soil. It can consist of consolidated rocks, and it can also include unconsolidated deposits such as river alluvium, lake or marine sediments, glacial tills, loess (silt-sized, wind-deposited particles), volcanic ash, and organic matter (such as accumulations in swamps or bogs). Parent materials influence soil formation through their mineralogical composition, their texture, and their stratification (occurrence in layers). Dark-coloured ferromagnesian (iron- and magnesium-containing) rocks, for example, can produce soils with a high content of iron compounds and of clay minerals in the kaolin or smectite groups, whereas light-coloured siliceous (silica-containing) rocks tend to produce soils that are low in iron compounds and that contain clay minerals in the illite or vermiculite groups. The coarse texture of granitic rocks leads to a coarse, loamy soil texture and promotes the development of E horizons (the leached lower regions of the topmost soil layer). The fine texture of basaltic rocks, on the other hand, yields soils with a loam or clay-loam texture and hinders the development of E horizons. Because water percolates to greater depths and drains more easily through soils with coarse texture, clearly defined E horizons tend to develop more fully on coarse parent material.

Leptosol soil profile from Switzerland, showing a typically shallow
surface horizon with little evidence of soil formation.

In theory, parent material is either freshly exposed solid matter (for example, volcanic ash immediately after ejection) or deep-lying geologic material that is isolated from atmospheric water and organisms. In practice, parent materials can be deposited continually by wind, water, or volcanoes and can be altered from their initial, isolated state, thereby making identification difficult. If a single parent material can be established for an entire soil profile, the soil is termed monogenetic; otherwise, it is polygenetic. An example of polygenetic soils are soils that form on sedimentary rocks or unconsolidated water- or wind-deposited materials. These so-called stratified parent materials can yield soils with intermixed geologic layering and soil horizons—as occurs in southeastern England, where soils forming atop chalk bedrock layers are themselves overlain by soil layers formed on both loess and clay materials that have been modified by dissolution of the chalk below.

Adjacent soils frequently exhibit different profile characteristics because of differing parent materials. These differing soil areas are called lithosequences, and they fall into two general types. Continuous lithosequences have parent materials whose properties vary gradually along a transect, the prototypical example being soils formed on loess deposits at increasing distances downwind from their alluvial source. Areas of such deposits in the central United States or China show systematic decreases in particle size and rate of deposition with increasing distance from the source. As a result, they also show increases in clay content and in the extent of profile development from weathering of the loess particles.

Andosol soil profile from Italy, showing a dark-coloured surface
horizon derived from volcanic parent material.

By contrast, discontinuous lithosequences arise from abrupt changes in parent material. A simple example might be one soil formed on schist (a silicate-containing metamorphic rock rich in mica) juxtaposed with a soil formed on serpentine (a ferromagnesian metamorphic rock rich in olivine). More subtle discontinuous lithosequences, such as those on glacial tills, show systematic variation of mineralogical composition or of texture in unconsolidated parent materials.

Topography

Topography, when considered as a soil-forming factor, includes the following: the geologic structural characteristics of elevation above mean sea level, aspect (the compass orientation of a landform), slope configuration (i.e., either convex or concave), and relative position on a slope (that is, from the toe to the summit). Topography influences the way the hydrologic cycle affects earth material, principally with respect to runoff processes and evapotranspiration. Precipitation may run off the land surface, causing soil erosion, or it may percolate into soil profiles and become part of subsurface runoff, which eventually makes its way into the stream system. Erosive runoff is most likely on a convex slope just below the summit, whereas lateral subsurface runoff tends to cause an accumulation of soluble or suspended matter near the toeslope. The conversion of precipitation into evapotranspiration is favoured by lower elevation and an equatorially facing aspect.

Adjacent soils that show differing profile characteristics reflecting the influence of local topography are called toposequences. As a general rule, soil profiles on the convex upper slopes in a toposequence are more shallow and have less distinct subsurface horizons than soils at the summit or on lower, concave-upward slopes. Organic matter content tends to increase from the summit down to the toeslope, as do clay content and the concentrations of soluble compounds.

Often the dominant effect of topography is on subsurface runoff (or drainage). In humid temperate regions, well-drained soil profiles near a summit can have thick E horizons (the leached layers) overlying well-developed clay-rich Bt horizons, while poorly drained profiles near a toeslope can have thick A horizons overlying extensive Bg horizons (lower layers whose pale colour signals stagnation under water-saturated conditions). In humid tropical regions with dry seasons, these profile characteristics give way to less distinct horizons, with accumulation of silica, manganese, and iron near the toeslope, whereas in semiarid regions soils near the toeslope have accumulations of the soluble salts sodium chloride or calcium sulfate.

These general conclusions are tempered by the fact that topography is susceptible to great changes over time. Soil erosion by water or wind removes A horizons and exposes B horizons to weathering. Major portions of entire soil profiles can move downslope suddenly by the combined action of water and gravity. Catastrophic natural events, such as volcanic eruptions, earthquakes, and devastating storms, can have obvious consequences for the instability of geomorphologic patterns.

Climate

The term climate in pedology refers to the characteristics of weather as they evolve over time scales longer than those necessary for soil properties to develop. These characteristics include precipitation, temperature, and storm patterns—both their averages and their variation.

Climate influences soil formation primarily through effects of water and solar energy. Water is the solvent in which chemical reactions take place in the soil, and it is essential to the life cycles of soil organisms. Water is also the principal medium for the erosive or percolative transport of solid particles. The rates at which these water-mediated processes take place are controlled by the amount of energy available from the sun.

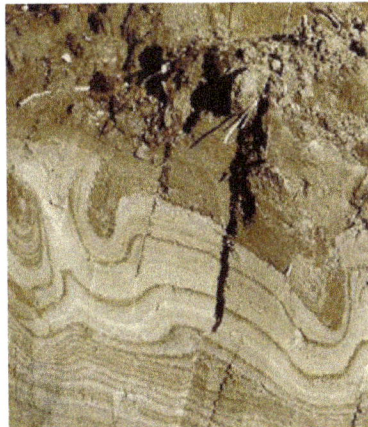

Cryosol soil profile from Canada showing patterned deformations caused by cycles of freeze and thaw.

On a global scale, the integrated effects of climate can readily be seen along a transect from pole to Equator. As one proceeds from the pole to cool tundra or forested regions, polar desert soils give way to intensively leached soils such as the Podzols (Spodosols) that exhibit an eye-catching, ash-coloured E horizon indicative of humid, boreal climates. Farther into temperate zones, organic matter accumulates in soils as climates become warmer, and eventually lime (calcium carbonate) also begins to accumulate closer to the top of the soil profile as evapotranspiration increases. Arid subtropical climate then follows, with desert soils that are low in organic matter and enriched in soluble salts. As the climate again becomes humid close to the Equator, high temperature combines with high precipitation to create red and yellow tropical soils, whose colours reveal the prevalence of residual iron oxide minerals that are resistant to leaching losses because of their low solubility.

On a continental scale, a transect taken across the central United States from east to west shows the effects of increasing evapotranspiration. First, soils that exhibit E horizons appear, followed by soils high in organic matter. These give way to soils with accumulations of lime and ultimately to desert soils with soluble salt efflorescence (powdery crust) near the surface.

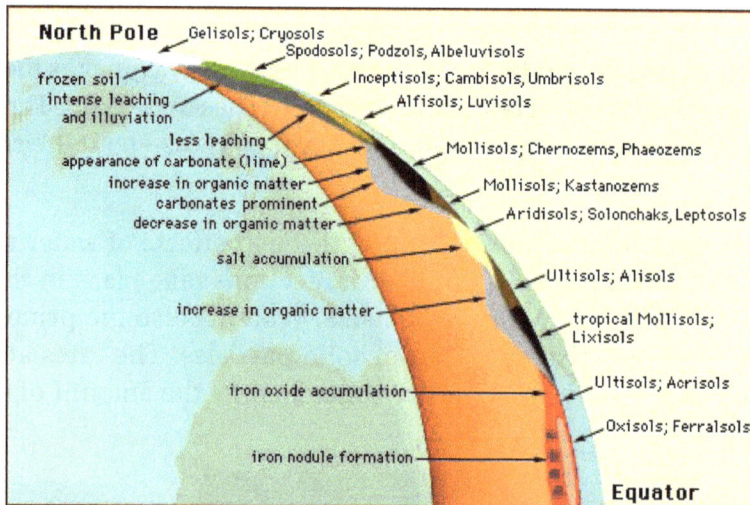

The effects of climate on soil worldwide A schematic cross section of the Earth illustrates the variation in soil types arising from differences in climate from the North Pole to the Equator.

On a regional scale, variations in climate also can influence soil properties significantly, resulting in a contiguous array of soils called a climosequence. One typical climosequence occurs along a 1,000-km (600-mile) north-south transect through the foothills of the Cascade and Sierra Nevada mountains in California. There soils that have formed on landscapes of similar topography vary continuously in their profile characteristics with variations in annual precipitation. Soils formed at the dry southern end of the transect are shallow and rocky, whereas those at the humid northern end show well-developed B horizons and reddish colour. Clay mineralogy in the upper 20 cm (8 inches) of these soils also responds to the increase in precipitation, shifting from the smectite group to the mixed vermiculite or illite group/kaolin group and finally to the kaolin group alone. These changes result primarily from increasing loss of silica and soluble metals as soil leaching extends deeper with increasing rainfall. In addition, soil acidity and organic matter content increase, while readily soluble forms of calcium (important to plant growth and soil aggregation) decrease, with increasing precipitation.

In principle, soil profile characteristics that are closely linked to climate can in turn be interpreted as climatic indicators. For example, a soil profile with two well-defined zones of lime accumulation, one shallow and one deep, may signal the existence of a past climate whose greater precipitation drove the lime layer deeper than the present climate is able to do.

Soils that formed in past environments different from the present and that are preserved (at least partially) at greater depth are known as paleosols. Some features of these soils can serve as climatic indicators, the most reliable being robust features such as horizons with hardened accumulations of relatively insoluble iron, manganese, or calcium minerals or layers with accumulations of strongly aggregated clay-size particles. Given a knowledge of the clay mineral in a suspected paleosol, and assuming the precipitation-clay mineralogy relationship, pedologists might be able to infer past climate. The precipitation level of a past climate might be inferred from an observation of the depth of lime-containing horizons in a paleosol. These potential applications of climatic relationships must be evaluated carefully in order to distinguish the effects of previously weathered parent material from those of purely climatic influence.

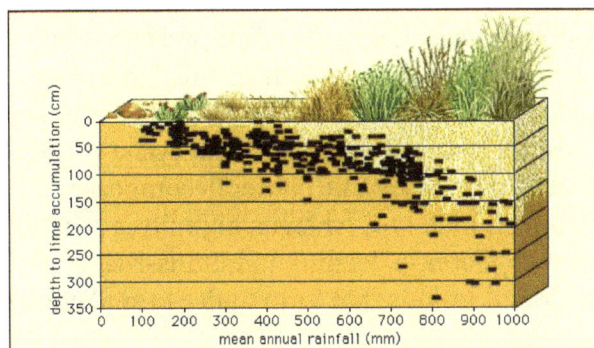

Depth to lime accumulation in relation to annual rainfall Lime ($CaCO_3$) deposits that can prevent the penetration of plant roots are found deeper in the soil profile in climates with higher mean annual rainfall than in climates where there is little water to transport the lime through the soil.

Organisms

The development of soils can be significantly affected by vegetation, animal inhabitants, and human populations. Any array of contiguous soils influenced by local flora and fauna is termed a biosequence. To return to the climosequence along the Cascade and Sierra Nevada ranges, the vegetation observed along this narrow foothill region varies from shrubs in the dry south to needle-leaved trees in the humid north, with extensive grasslands in between. In the middle of the precipitation range, transition zones occur in which small groves of needle-leaved trees are interspersed with grassland patches in an apparently random manner. These plant populations represent local flora largely selected by climate. The properties of the soils underlying these plants, however, exhibit differences that do not arise from climate, topography, or parent material but are an effect of the differing plant species. The soils under trees, for instance, are much more acidic and contain much less humus than those under grass, and nitrogen content is considerably greater in the grassland soil. These properties come directly from the type of litter produced by the two different kinds of vegetation.

Anthrosol soil profile from The Netherlands, showing surface layers made homogeneous through human activity.

An opportunity to examine biosequences is often presented by relatively young soils formed from an alluvial parent material. Soils of this kind lying beneath shrubs may be richer in humus and plant nutrients than similar soils found beneath needle-leaved trees. This variation results from differences in the cyclic processes of plant growth, litter production, and litter decay. Organic matter decomposers will feed on stored material in soil if litter production is low, whereas high litter production will permit soil stocks of organic matter to increase, leading to humus-rich A horizons as opposed to the leached E horizons found in soils that form under humid climatic conditions.

Human beings are also part of the biological influx that influences soil formation. Human influence can be as severe as wholesale removal or burial (by urbanization) of an entire soil profile, or it can be as subtle as a gradual modification of organic matter by agriculture or of soil structure by irrigation. The chemical and physical properties of soils critical to the growth of crops often are affected significantly by cultural practices. Among the problems created for agriculture by cultural practices themselves are loss of arable land, erosion, the buildup of salinity, and the depletion of organic matter.

Time

The soil-forming factors of parent material and topography are largely site-related (attributes of the terrain), whereas those of climate and organisms are largely flux-related (inputs from the surroundings). Time as a soil-forming factor is neither a property of the terrain nor a source of external stimulus. It is instead an abstract variable whose significance is solely as a marker of the evolution of soil characteristics. The conceptual independence of time from its four companion factors means simply that soil evolution can occur while site attributes and external inputs remain essentially unchanged.

Certain soil profile features can be interpreted as indicators of the passage of time. (A series of soil profiles whose features differ only as a result of age constitutes a chronosequence.) One example of a time-related feature is the humus content of the A horizon, which, for soils less than 10,000 years old, increases continually at a rate dependent on parent material, vegetation, and climate. Typically, this rate of increase slows after about 10,000 years, plant nutrients begin to leach away, and a significant decline in humus content is observed for soils whose age approaches one million years. (Agricultural practices can interrupt this trend, causing a gradual drop in stored humus by 25 percent or more. Correlated with these changes are economically important soil properties, such as nutrient supply and retention capability, acidity, and aeration.)

The accumulation of clay and lime in soil profiles as a result of their translocation downward is also an indication of aging. For example, older soils that have formed on calcium containing loess deposits have better-developed E and Bt horizons (as well as thinner A horizons) than younger soils forming on these deposits. Similarly, soils in a chronosequence developed on alluvium can exhibit a clayey hardpan after 100,000 years or so. Soils also tend to redden in colour as they age, irrespective of climatic conditions, reflecting the persistence of poorly crystalline or crystalline oxide minerals containing Fe^{3+}. Indeed, the dominant mineralogy of the clay-size particles in soils is itself a reliable indicator of soil age. Any particular sequence of predominant clay mineralogy found in a soil is known collectively as the set of Jackson-Sherman weathering stages. Each downward increment corresponds to increasing mineral residence time, both among and within the three principal stages (early, intermediate, and advanced).

	Characteristic minerals in soil clay fraction	Characteristic chemical and physical conditions of soil	Characteristic soil profile features
Early stage	Gypsum, carbonates, olivine/pyroxene/ amphibole, Fe^{2+}-bearing micas, and feldspars.	Low water and humus content, limited leaching, reducing environments, and limited time for weathering.	Minimally weathered soils all over the world, though mainly in arid regions where low rainfall keeps weathering to a minimum.
Intermediate stage	Quartz, mica/illite, vermiculite/chlorite, and smectite.	Retention of Na, K, Ca, Mg, Fe^{2+}, and silica; alkalinity and ineffective leaching; igneous rock rich in Ca, Mg, and Fe^{2+} but no Fe^{2+}oxides; easily hydrolyzed silicates; and transport of silica into the weathering zone.	Soils of temperate regions developed under grass or trees—i.e., the major agricultural soils of the world.
Advanced stage	Kaolin, aluminum and iron oxides, and titanium oxides.	Removal of Na, K, Ca, Mg, Fe^{2+}, and silica; effective leaching by fresh water; low pH; and dispersion of silica.	Intensely weathered soils of the humid tropics, frequently characterized by acidity and low fertility.

The early stage of weathering is recognized through the dominance of sulfates, carbonates, and primary silicates, other than quartz and muscovite, in the soil clay fraction. These minerals can survive only if soils remain very dry, very cold, or very wet most of the time—that is, if they have limited exposure to water, air, or solar energy. The intermediate stage features quartz, muscovite, and secondary aluminosilicates prominently in the clay fraction. These minerals can survive under leaching conditions that do not deplete silica and metals, such as calcium or magnesium, and that do not result in the complete oxidation of Fe^{2+}, which is then incorporated into illite and smectite clays. The advanced stage, on the other hand, is associated with intensive leaching and strong oxidizing conditions, such that only oxides of aluminum, Fe^{3+}, and titanium remain. Kaolin will be a dominant clay mineral group only if the removal of silica by leaching is not complete or if there is an encroachment of silica-rich waters—as can occur, for example, when water percolating through a soil profile at the upper part of a toposequence moves downslope into a soil profile at the lower part.

Implicit is the conclusion that more time is required to form soils featuring the more persistent minerals in the clay fraction. This conclusion is borne out by careful field studies worldwide in which the rate of soil horizon formation is determined. Soils whose clay mineralogy falls in the early stage require less than a decade to develop a centimetre (0.4 inch) of horizon thickness. Soils with dominant clay minerals in the intermediate stage do this in less than a century, whereas soils with dominant clay minerals in the advanced stage need several hundred years to form a centimetre of solum.

Types of Soil

The soil is basically classified into four types:

- Sandy soil.

- Silt Soil.

- Clay Soil.

- Loamy Soil.

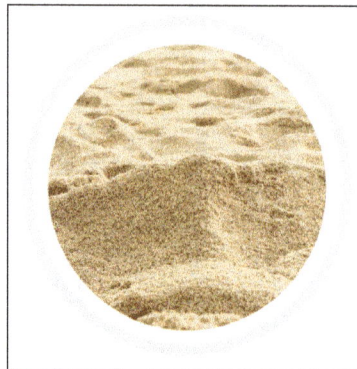

Sandy Soil

The first type of soil is sand. It consists of small particles of weathered rock. Sandy soils are one of the poorest types of soil for growing plants because it has very low nutrients and poor in holding water, which makes it hard for the plant's roots to absorb water. This type of soil is very good for the drainage system. Sandy soil is usually formed by the breakdown or fragmentation of rocks like granite, limestone, and quartz.

Silt Soil

Silt, which is known to have much smaller particles compared to the sandy soil and is made up of rock and other mineral particles which are smaller than sand and larger than clay. It is the smooth and quite fine quality of the soil that holds water better than sand. Silt is easily transported by moving currents and it is mainly found near the river, lake, and other water bodies. The slit soil is more fertile compared to the other three types of soil. Therefore it is also used in agricultural practices to improve soil fertility.

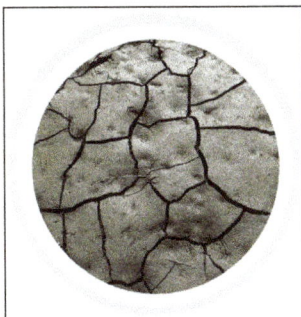

Clay Soil

Clay is the smallest particles amongst the other two types of soil. The particles in this soil are tightly packed together with each other with very little or no airspace. This soil has very good water storage qualities and making hard for moisture and air to penetrate into it. It is very sticky to the touch when wet, but smooth when dried. Clay is the densest and heaviest type of soil which do not drain well or provide space for plant roots to flourish.

Loamy Soil

Loam is the fourth type of soil. It is a combination of sand, silt, and clay such that the beneficial properties from each is included. For instance, it has the ability to retain moisture and nutrients, hence, it is more suitable for farming. This soil is also referred to as an agricultural soil as it includes an equilibrium of all three types of soil materials being sandy, clay, and silt and it also happens to have hummus. Apart from these, it also has higher calcium and pH levels because of its inorganic origins.

2

Chemistry of Soil

The study of the chemical characteristics of soil is termed as soil chemistry. Some of the major areas of study within this field are soil pH, soil reaction, soil acidity and soil alkalinity. The topics elaborated in this chapter will help in gaining a better perspective about these focus areas of soil chemistry.

SOIL CHEMISTRY

Soil chemistry can be considered as the natural chemical composition of a given soil. This natural chemical composition of a soil is a function of that soil's parent material. In many areas of the world soil is formed in place and derived directly from the weathering and degradation of rocks. When soil is derived from rocks, its soil chemistry is a direct reflection of the rocks' chemistry, including the minerals found in the rocks. There are different forms of rocks, including the major classifications of sedimentary, metamorphic, and igneous. Within each of these major separations there are different categories of rocks. Thus, rocks are a collection of different minerals. Soils derived from these different rocks will have different chemical complexes. For example, soils derived from sandstone will have a limited array of different chemicals, whereas soils derived from igneous rock might have a greater diversity of minerals and chemicals. However, not all soils are formed in situ from the rocks that exist in place.

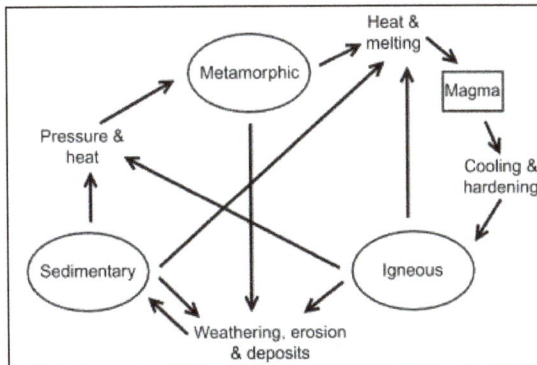

Simplified rock cycle and major rock material available to form soil. Note that rocks vary in their mineral and chemical composition, which in turn affects the type of soil derived from these rocks.

Soil parent materials range from glacial drift (glacial till), to water deposited (fluvial) material, and mass wasting of primary rock or other deposit caused by gravitational forces (colluvium), among others. Soil that is derived from glacial drift or any of these other sources will likely have a very complex chemical composition reflecting the many rock types and other materials entrained in the parent material. Thus, a natural soil's chemical composition is a reflection of its source materials. While soil will inherit a given mineral chemical composition, the organic composition is derived from living organisms (e.g., animals, plants, fungi). It is this organic matter that is often considered the key to good soil health.

Natural soil chemistry can be changed by various natural forces such as leaching of chemical elements by water moving through the soil, chemical reactions, and biological activity. However, soil chemistry can also be altered by human impact from various land uses, including farming. It is a given that soil chemistry can, and will, be changed by natural forces, and one might consider these changes negative or positive impacts depending on the intended future use of the soil. Similar to natural changes, management of soils by humans can result in positive or negative changes to soil chemistry, such as increases in soil organic matter or soil erosion because of aggressive tillage practices. For example, fertilizer and manure applications to a silt loam soil increased the percentage of water-stable aggregates of a silt loam in Romania because of increases in soil organic matter. Other management practices, such as inversion tillage, can increase the erodibility of soil because of soil aggregate disturbance.

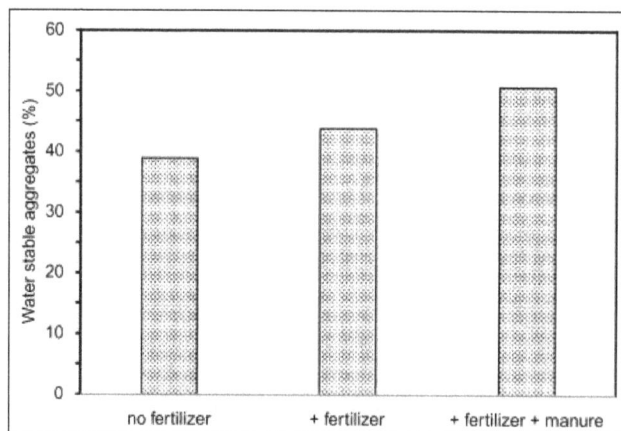

Impact of fertilizer and manure application to soybean-wheat-corn-barley rotation on water-stable aggregates of a silt loam from northeastern Romania.

The greatest negative effect of CAPS on soil chemistry is not only the intense use and introduction of herbicides and insecticides in the soil ecosystem, but the lack of crop rotations and cover crops which are known for adding organic carbon (C) and biological diversity to the soil system. The tillage systems used in most CAPS result in substantial soil disturbance and generally a negative impact on soil organic matter. Aggressively tilling the soil leads to a considerable incorporation of air, which causes a spike in microbial activity and subsequently accelerates soil organic matter breakdown. Since soil organic matter is key to good soil health, any reductions in soil organic matter will obviously reduce soil health.

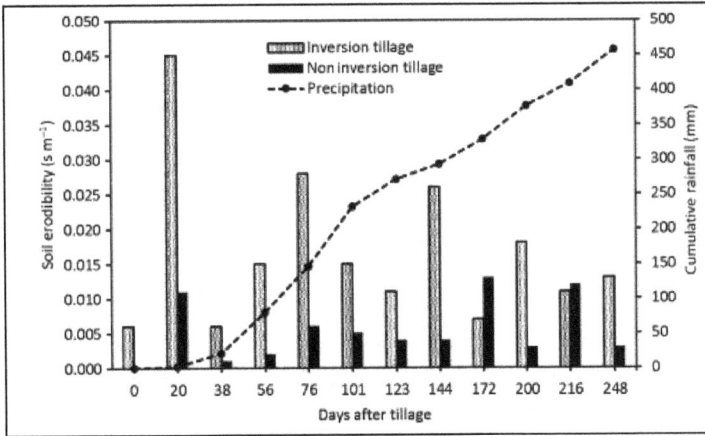

The concentrated flow soil erodibility factor increases after inversion tillage compared to noninversion tillage because of greater soil aggregate disturbance of a Belgian silt loam.

It has been shown that excessive introduction of certain N fertilizers into the soil ecosystem results in soil acidification. This added acidity further degrades soil by affecting soil primary particles, namely clay minerals. This was demonstrated by Barak et al. for glacially deposited soils in south central Wisconsin, United States, where they found a significant increase in soil acidity caused decreases in cation exchange capacity (CEC), base saturation, and exchangeable calcium (Ca) and magnesium (Mg). Some of the degradations in soil properties were considerable, e.g., clays in this soil showed a 16% loss of CEC. These degradations came from application of anhydrous ammonia as a N fertilizer in long-term fertility experiments. These results highlight the impact that CAPS management practices can have on soil health and productivity.

In addition to fertilizer applications, soil chemistry is altered by pesticide applications. When pesticides, especially insecticides are applied to the soil they sometimes change the soil chemical environment to the detriment of soil biological populations. Good soil quality is characterized as having a vast range of soil microorganisms. While soil-applied insecticides have been proven to generally kill target insects, they will also destroy "beneficial" organisms.

When excessive pesticides are used, which is generally the case in CAPS, not only do we see degradation of the soil ecosystem, but the underlying groundwater system often becomes polluted. It was previously noted that soil chemistry is altered by leaching of chemical substances by water. Chemical leaching is not limited to natural soil chemical elements but synthetic chemicals applied to soil, such as pesticides and fertilizers, leach as well. This is especially true for sandy soils and soils with little organic matter content. Most pesticides are strongly retained by soil organic matter and clay minerals. Sandy soils are often low in organic matter and CEC, thus providing ideal conditions for pesticides and nitrate from organic and synthetic N fertilizers to leach. In addition, sandy soils tend to be well drained, thus water can move rapidly through them. Groundwater contamination is often reported in areas where the water table is close to the soil surface.

SOIL pH

Soil pH is a measure of the acidity or basicity (alkalinity) of a soil. pH is defined as the negative logarithm (base 10) of the activity of hydronium ions (H^+ or, more precisely, $H_3O^+_{aq}$) in a solution. In soils, it is measured in a slurry of soil mixed with water (or a salt solution, such as 0.01 M $CaCl_2$), and normally falls between 3 and 10, with 7 being neutral. Acid soils have a pH below 7 and alkaline soils have a pH above 7. Ultra-acidic soils (pH < 3.5) and very strongly alkaline soils (pH > 9) are rare.

Soil pH is considered a master variable in soils as it affects many chemical processes. It specifically affects plant nutrient availability by controlling the chemical forms of the different nutrients and influencing the chemical reactions they undergo. The optimum pH range for most plants is between 5.5 and 7.5; however, many plants have adapted to thrive at pH values outside this range.

Classification of Soil pH Ranges

The United States Department of Agriculture Natural Resources Conservation Service classifies soil pH ranges as follows:

Denomination	pH range
Ultra acidic	< 3.5
Extremely acidic	3.5–4.4
Very strongly acidic	4.5–5.0
Strongly acidic	5.1–5.5
Moderately acidic	5.6–6.0
Slightly acidic	6.1–6.5
Neutral	6.6–7.3
Slightly alkaline	7.4–7.8
Moderately alkaline	7.9–8.4
Strongly alkaline	8.5–9.0
Very strongly alkaline	> 9.0

Determining pH

Methods of determining pH include:

- Observation of soil profile: Certain profile characteristics can be indicators of either acid, saline, or sodic conditions. Examples are:

  ◦ Poor incorporation of the organic surface layer with the underlying mineral layer – this can indicate strongly acidic soils;

  ◦ The classic podzol horizon sequence, since podzols are strongly acidic: in these soils, a pale eluvial (E) horizon lies under the organic surface layer and overlies a dark B horizon;

  ◦ Presence of a caliche layer indicates the presence of calcium carbonates, which are present in alkaline conditions;

  ◦ Columnar structure can be an indicator of sodic condition.

- Observation of predominant flora. Calcifuge plants (those that prefer an acidic soil) include *Erica*, *Rhododendron* and nearly all other Ericaceae species, many birch (*Betula*), foxglove (*Digitalis*), gorse (*Ulex* spp.), and Scots Pine (*Pinus sylvestris*). Calcicole (lime loving) plants include ash trees (*Fraxinus* spp.), honeysuckle (*Lonicera*), *Buddleja*, dogwoods (*Cornus* spp.), lilac (*Syringa*) and *Clematis* species.

- Use of an inexpensive pH testing kit, where in a small sample of soil is mixed with indicator solution which changes colour according to the acidity.

- Use of litmus paper. A small sample of soil is mixed with distilled water, into which a strip of litmus paper is inserted. If the soil is acidic the paper turns red, if basic, blue.

- Use of a commercially available electronic pH meter, in which a glass or solid-state electrode is inserted into moistened soil or a mixture (suspension) of soil and water; the pH is usually read on a digital display screen.

- Recently, spectrophotometric methods have been developed to measure soil pH involving addition of an indicator dye to the soil extract. These compared well to glass electrode measurements but offer substantial advantages such as lack of drift, liquid junction and suspension effects.

Precise, repeatable measures of soil pH are required for scientific research and monitoring. This generally entails laboratory analysis using a standard protocol.

Factors affecting Soil pH

The pH of a natural soil depends on the mineral composition of the parent material of the soil, and the weathering reactions undergone by that parent material. In warm, humid environments, soil acidification occurs over time as the products of weathering are leached by water moving laterally or downwards through the soil. In dry climates, however, soil weathering and leaching are less intense and soil pH is often neutral or alkaline.

Sources of Acidity

Many processes contribute to soil acidification. These include:

- Rainfall: Acid soils are most often found in areas of high rainfall. Rainwater has a slightly acidic pH (usually about 5.7) due to a reaction with CO_2 in the atmosphere that forms carbonic acid. When this water flows through soil it results in the leaching of basic cations from the soil as bicarbonates; this increases the percentage of Al^{3+} and H^+ relative to other cations.

- Root respiration and decomposition of organic matter by microorganisms releases CO_2 which increases the carbonic acid (H_2CO_3) concentration and subsequent leaching.

- Plant growth: Plants take up nutrients in the form of ions (e.g. NO_3^-, NH_4^+, Ca^{2+}, $H_2PO_4^-$), and they often take up more cations than anions. However plants must maintain a neutral charge in their roots. In order to compensate for the extra positive charge, they will release H^+ ions from the root. Some plants also exude organic acids into the soil to acidify the zone around their roots to help solubilize metal nutrients that are insoluble at neutral pH, such as iron (Fe).

- Fertilizer use: Ammonium (NH_4^+) fertilizers react in the soil by the process of nitrification to form nitrate (NO_3^-), and in the process release H^+ ions.

- Acid rain: The burning of fossil fuels releases oxides of sulfur and nitrogen into the atmosphere. These react with water in the atmosphere to form sulfuric and nitric acid in rain.

- Oxidative weathering: Oxidation of some primary minerals, especially sulfides and those containing Fe^{2+}, generate acidity. This process is often accelerated by human activity:

  ◦ Mine spoil: Severely acidic conditions can form in soils near some mine spoils due to the oxidation of pyrite.

  ◦ Acid sulfate soils formed naturally in waterlogged coastal and estuarine environments can become highly acidic when drained or excavated.

Sources of Alkalinity

Total soil alkalinity increases with:

- Weathering of silicate, aluminosilicate and carbonate minerals containing Na^+, Ca^{2+}, Mg^{2+} and K^+.

- Addition of silicate, aluminosilicate and carbonate minerals to soils; this may happen by deposition of material eroded elsewhere by wind or water, or by mixing of the soil with less weathered material (such as the addition of limestone to acid soils).

- Addition of water containing dissolved bicarbonates (as occurs when irrigating with high-bicarbonate waters).

The accumulation of alkalinity in a soil (as carbonates and bicarbonates of Na, K, Ca and Mg) occurs when there is insufficient water flowing through the soils to leach soluble salts. This may be due to arid conditions, or poor internal soil drainage; in these situations most of the water that enters the soil is transpired (taken up by plants) or evaporates, rather than flowing through the soil.

The soil pH usually increases when the total alkalinity increases, but the balance of the added cations also has a marked effect on the soil pH. For example, increasing the amount of sodium in an alkaline soil tends to induce dissolution of calcium carbonate, which increases the pH. Calcareous soils may vary in pH from 7.0 to 9.5, depending on the degree to which Ca^{2+} or Na^+ dominate the soluble cations.

Effect of Soil pH on Plant Growth

Acid Soils

Plants grown in acid soils can experience a variety of stresses including aluminium (Al), hydrogen (H), and/or manganese (Mn) toxicity, as well as nutrient deficiencies of calcium (Ca) and magnesium (Mg).

Aluminium toxicity is the most widespread problem in acid soils. Aluminium is present in all soils, but dissolved Al^{3+} is toxic to plants; Al^{3+} is most soluble at low pH; above pH 5.0, there is little Al in soluble form in most soils. Aluminium is not a plant nutrient, and as such, is not actively taken up by the plants, but enters plant roots passively through osmosis. Aluminium inhibits root growth; lateral roots and root tips become thickened and roots lack fine branching; root tips may turn brown. In the root, the initial effect of Al^{3+} is the inhibition of the expansion of the cells of the rhizodermis, leading to their rupture; thereafter it is known to interfere with many physiological processes including the uptake and transport of calcium and other essential nutrients, cell division, cell wall formation, and enzyme activity.

Proton (H^+ ion) stress can also limit plant growth. The proton pump, H^+-ATPase, of the plasmalemma of root cells works to maintain the near-neutral pH of their cytoplasm. A high proton activity (pH within the range 3.0–4.0 for most plant species) in the external growth medium overcomes the capacity of the cell to maintain the cytoplasmic pH and growth shuts down.

In soils with a high content of manganese-containing minerals, Mn toxicity can become a problem at pH 5.6 and lower. Manganese, like aluminium, becomes increasingly soluble as pH drops, and Mn toxicity symptoms can be seen at pH levels below 5.6. Manganese is an essential plant nutrient, so plants transport Mn into leaves. Classic symptoms of Mn toxicity are crinkling or cupping of leaves.

Nutrient Availability in Relation to Soil pH

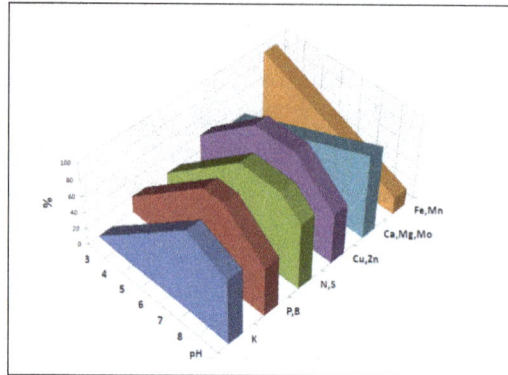

Soil pH affects the availability of some plant nutrients. Aluminium toxicity has direct effects on plant growth; however, by limiting root growth, it also reduces the availability of plant nutrients. Because roots are damaged, nutrient uptake is reduced, and deficiencies of the macronutrients (nitrogen, phosphorus, potassium, calcium and magnesium) are frequently encountered in very strongly acidic to ultra-acidic soils (pH<5.0).

Molybdenum availability is increased at higher pH; this is because the molybdate ion is more strongly sorbed by clay particles at lower pH.

Zinc, iron, copper and manganese show decreased availability at higher pH (increased sorbtion at higher pH).

The effect of pH on phosphorus availability varies considerably, depending on soil conditions and the crop in question. The prevailing view in the 1940s and 1950s was that P availability was maximized near neutrality (soil pH 6.5–7.5), and decreased at higher and lower pH. Interactions of phosphorus with pH in the moderately to slightly acidic range (pH 5.5–6.5) are, however, far more complex than is suggested by this view. Laboratory tests, glasshouse trials and field trials have indicated that increases in pH within this range may increase, decrease, or have no effect on P availability to plants.

Water Availability in Relation to Soil pH

Strongly alkaline soils are sodic and dispersive, with slow infiltration, low hydraulic conductivity and poor available water capacity. Plant growth is severely restricted because aeration is poor when the soil is wet; in dry conditions, plant-available water is rapidly depleted and the soils become hard and cloddy (high soil strength).

Many strongly acidic soils, on the other hand, have strong aggregation, good internal drainage, and good water-holding characteristics. However, for many plant species, aluminium toxicity severely limits root growth, and moisture stress can occur even when the soil is relatively moist.

Plant pH Preferences

In general terms, different plant species are adapted to soils of different pH ranges. For many species, the suitable soil pH range is fairly well known.

However, a plant may be intolerant of a particular pH in some soils as a result of a particular mechanism, and that mechanism may not apply in other soils. For example, a soil low in molybdenum may not be suitable for soybean plants at pH 5.5, but soils with sufficient molybdenum allow optimal growth at that pH. Similarly, some calcifuges (plants intolerant of high-pH soils) can tolerate calcareous soils if sufficient phosphorus is supplied. Another confounding factor is that different varieties of the same species often have different suitable soil pH ranges. Plant breeders can use this to breed varieties that can tolerate conditions that are otherwise considered unsuitable for that species – examples are projects to breed aluminium-tolerant and manganese-tolerant varieties of cereal crops for food production in strongly acidic soils.

The table below gives suitable soil pH ranges for some widely cultivated plants as found in the *USDA PLANTS Database*. Some species (like *Pinus radiata* and *Opuntia ficus-indica*) tolerate only a narrow range in soil pH, whereas others (such as *Vetiveria zizanioides*) tolerate a very wide pH range.

Scientific name	Common name	pH (minimum)	pH (maximum)
Vetiveria zizanioides	Vetivergrass	3.0	8.0
Pinus rigida	Pitch pine	3.5	5.1
Rubus chamaemorus	Cloudberry	4.0	5.2
Ananas comosus	Pineapple	4.0	6.0
Coffea arabica	Arabian coffee	4.0	7.5
Rhododendron arborescens	Smooth azalea	4.2	5.7
Pinus radiata	Monterey pine	4.5	5.2
Carya illinoinensis	Pecan	4.5	7.5
Tamarindus indica	Tamarind	4.5	8.0
Vaccinium corymbosum	Highbush blueberry	4.7	7.5
Manihot esculenta	Cassava	5.0	5.5
Morus alba	White mulberry	5.0	7.0
Malus	Apple	5.0	7.5
Pinus sylvestris	Scots pine	5.0	7.5

Scientific name	Common name	pH (minimum)	pH (maximum)
Carica papaya	Papaya	5.0	8.0
Cajanus cajan	Pigeonpea	5.0	8.3
Pyrus communis	Common pear	5.2	6.7
Solanum lycopersicum	Garden tomato	5.5	7.0
Psidium guajava	Guava	5.5	7.0
Nerium oleander	Oleander	5.5	7.8
Punica granatum	Pomegranate	6.0	6.9
Viola sororia	Common blue violet	6.0	7.8
Caragana arborescens	Siberian peashrub	6.0	9.0
Cotoneaster integerrimus	Cotoneaster	6.8	8.7
Opuntia ficus-indica	Barbary fig (pricklypear)	7.0	8.5

Changing Soil pH

Increasing pH of Acidic Soil

Finely ground agricultural lime is often applied to acid soils to increase soil pH (liming). The amount of limestone or chalk needed to change pH is determined by the mesh size of the lime (how finely it is ground) and the buffering capacity of the soil. A high mesh size (60 mesh = 0.25 mm; 100 mesh = 0.149 mm) indicates a finely ground lime that will react quickly with soil acidity. The buffering capacity of a soil depends on the clay content of the soil, the type of clay, and the amount of organic matter present, and may be related to the soil cation exchange capacity. Soils with high clay content will have a higher buffering capacity than soils with little clay, and soils with high organic matter will have a higher buffering capacity than those with low organic matter. Soils with higher buffering capacity require a greater amount of lime to achieve an equivalent change in pH.

Amendments other than agricultural lime that can be used to increase the pH of soil include wood ash, industrial calcium oxide (burnt lime), magnesium oxide, basic slag (calcium silicate), and oyster shells. These products increase the pH of soils through various acid-base reactions. Calcium silicate neutralizes active acidity in the soil by reacting with H^+ ions to form monosilicic acid (H_4SiO_4), a neutral solute.

Decreasing the pH of Alkaline Soil

The pH of an alkaline soil can be reduced by adding acidifying agents or acidic organic materials. Elemental sulfur (90–99% S) has been used at application rates of 300–500 kg/ha – it slowly oxidizes in soil to form sulfuric acid. Acidifying fertilizers, such as ammonium sulfate, ammonium nitrate and urea, can help to reduce the pH of a soil because ammonium oxidises to form nitric acid. Acidifying organic materials include peat or sphagnum peat moss.

However, in high-pH soils with a high calcium carbonate content (more than 2%), it can be very costly and/or ineffective to attempt to reduce the pH with acids. In such cases, it is often more efficient to add phosphorus, iron, manganese, copper and/or zinc instead, because deficiencies of these nutrients are the most common reasons for poor plant growth in calcareous soils.

Measurement of Soil pH

There are many different ways of measuring the pH of soil. Some use a saturated paste extract, others use a 1:5 dilution of soil: water, and then take a pH measurement on the resulting solution with a laboratory meter. Others use the 1:5 dilution, but instead of water they use a dilute Calcium Chloride (CaCl2) solution. As a rough guide, the pH in CaCl2 is usually 0.8 pH lower than in water, but can be as much as 2.0 pH units lower on grey sands. Ensure that you check the method used to measure you soil pH so that you are comparing similar methods.

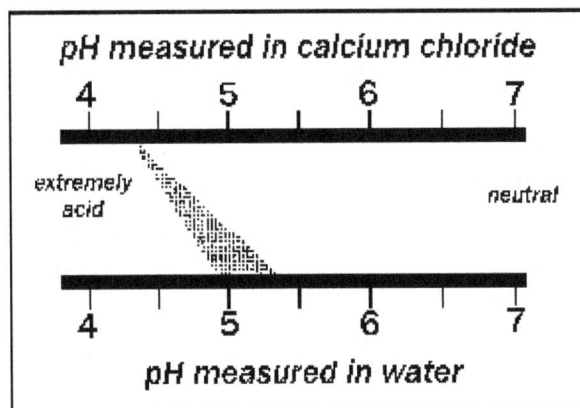

Using a Calcium Chloride Solution for Soil Test

Using a dilute CaCl^2 solution will probably give more consistent results than using rain-water or diluted water. When the soil is diluted with water , most of the H$^+$ ions tend to remain attracted to the soil particles and are not released into the soil solution. The addition of small amounts of calcium chloride provides Ca^{2+} ions to replace some of the H$^+$ ions on the soil particles, forcing the hydrogen ions into the solution and making their concentration in the bulk solution closer to that found in the field. The pH measured in CaCl^2 is almost always lower than pH of the same soil measured in water due to the higher concentration of H$^+$. The procedure gives a value similar to that for natural soil solution because the soil solution also contains dissolved Ca^{2+} and other ions.

Make up a dilute CaCl^2 solution with distilled/deionised water to use as you need it. Ready made calcium chloride solutions often do not keep for very long, so buy CaCl^2 and make the solution yourself. The calcium chloride is usually the dehydrate form (water is attached to the crystals - it will have this written on the label of the container

- CaCl².2H₂O). Dissolve approximately 7.5g of the salt in 5 litres of distilled/deionised water. If you are using a form of calcium chloride crystals without any water attached (CaCl²), dissolve about 5.5g of the salt in 5 litres. There is no need to be very accurate in your weighing as small errors will not effect the results.

Methods for Soil pH

Use the spoon to weigh out about 10g (to the nearest half gram), of your soil into the container. Add 50ml of distilled water to the soil. Any rough measurement ensuring a 1:5 diluted will suffice. Shake the container for about 2-3 minutes then allow the soil to settle for 2 minutes. If your soil has a high clay content and you require a very accurate result, it may not be necessary to filter the suspension. If filtering is not required, measure the pH value on the water above the soil in the container. Ensure you get a steady reading on the digital readout. Always wash your containers out before testing your next sample.

Testing Soil pH using pH Tester

Measurement of pH in soil is very common as it affects the relative availability of soil nutrients. If the pH is not within an acceptable range, growth will be curtailed and erosion potential is increased. The ability of the soil to provide adequate nutrition to the plant depends on four factors:

- The amount of various essential elements present in the soil.

- Their forms of combination.

- The processes by which these elements become available to plants.

- The soil solution and its pH.

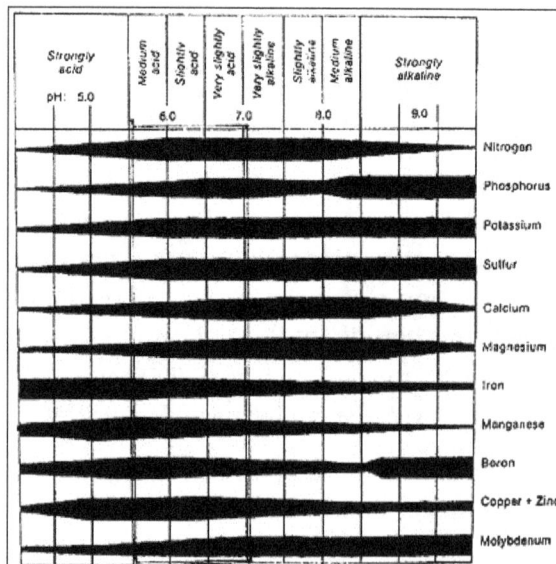

The effect of soil pH on availability of plant nutrients.

The amount of various elements present in the soil depends on the nature of the soil and on its organic matter content since it is a source of several nutrient elements. Soil nutrients exist both as complex, insoluble compounds and as simple forms usually soluble in soil water and readily available to plants. The complex forms must be broken down through decomposition to the simpler and more available forms in order to benefit the plant. The pH value is a measure of acidity or alkalinity. The effect of pH on availability of essential elements is shown.

Iron, manganese and zinc become less available as the pH is raised from 6.5 to 7.5 or 8.0. Molybdenum and phosphorus availability, on the other hand is affected in the opposite way, being greater at the higher pH levels. At very high pH values the bicarbonate ion (HCO_3) may be present in sufficient quantities to interfere with the normal uptake of other ions and thus detrimental to optimum growth. When inorganic salts are placed in a dilute solution they dissociate into electrically charged units called ions. These ions are available to the plant from the surface of the soil colloids and from salts in the soil solution. The positively charged ions (cations) such as potassium (K^+) and calcium (Ca_2^+) are mostly absorbed by the soil colloids, whereas the negatively charged ions (anions), such as chloride (Cl^-) and sulphate (SO_4^-) are found in the soil solution.

Soil Sample Preparation

1. Scoop up loose soil samples with a clean, dry plastic jar. Avoid touching the soil with your hands to prevent contaminating the sample.

2. Remove any stones and crush any clumps of soil to prevent breaking the delicate pHScan glass electrode bulb.

3. Fill up your sample soil up to 3/4 and add distilled water to the jar. Cap the jar tight and shake it vigorously a few times. Let the mixed sample stand for 5-10 minutes to dissolve the salts in the soil.

4. Prepare to log your test result in your data book for later reference.

5. Remove the caps of the jar and your pHScan. Dip the pHScan electrode into the wet soil slurry and turn the tester on. Take the reading when it stabilizes.

6. Press HOLD button to freeze the displayed pH measurement. Record the pH reading in your data book.

7. Press HOLD button again to release the reading.

8. Rinse your pH Scan tester in clean water between each use.

Calibrate the pH Scan tester using the instruction provided in the packaging box.

Soil pH Data from pHScan

The pH test value in this procedure is accurate to ±0.5 pH or better (usually ±0.2 pH). The soil sample preparation and test procedure is adapted from accepted laboratory methods. Soil pH testing in the field gives small differences between tests. Using the 0.1 pH resolution pHScan 1 minimizes these differences.

Most soil pH measurement cannot achieve ±0.1 pH accuracy, even with elaborate laboratory procedures and expensive pH instruments. Usually, soil pH data for many applications do not require testing for better accuracy than a few tenths of a pH or ±0.5 pH in some instances. pHScan 1 easily meets requirements for soil pH testing, and is very economical to use in places where many samples and tests are taken.

SOIL REACTION

One of the outstanding physiological characteristics of the soil solution is its reaction. Soil reaction influences many physical and chemical properties of soil. The growth and activity of plant and soil organisms depend on soil reaction and the factors associated with it.

There can be three types of soil reaction as follows:

Acidity

Soil acidity is common in regions where precipitation is high enough to leach appreciable amounts of exchangeable bases from the surface layers of the soil. The two adsorbed cations such as Hydrogen and Aluminium are largely responsible for soil acidity. The acid soil is generally found in humid region. The factors which will help in the release and removal of bases will help in the development of acidity of soil. If the hydrogen (H^+) ion becomes more than hydroxyl (OH^-) ion in the soil solution, the soil becomes acidic.

A highly acidic soil may have pH 4.5 and low calcium and magnesium, high solubility of iron, manganese, aluminium etc., but low availability of nitrogen and phosphorus. The activity of microorganism responsible for nitrification is adversely affected in acid soil. Generally limes are used for reclamation of acid soil.

Alkalinity

The soil that contains absorbed sodium to interfere with the growth of most crop plant is known as alkali soil. The amount of exchangeable sodium in great quantities in the soil makes the soil alkalinity. The sodium ion easily displaced the calcium ion from clay colloid and makes the sodium mixed clay particles. This sodium is converted into sodium hydroxide by hydrolysis as per the following reaction:

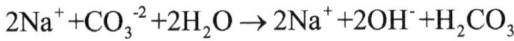

$$2Na^+ + CO_3^{-2} + 2H_2O \rightarrow 2Na^+ + 2OH^- + H_2CO_3$$

The OH^- ion thus formed increases the soil pH.

Neutrality

In those areas, where the soil contain hydrogen and hydroxyl ion almost in equal quantities, the soils are neutral in character.

Soil pH

pH is the negative logarithm of hydrogen ion concentration of soil solution and it is usually written as:

$$pH = \log \frac{1}{(H^+)} \text{ or } pH = -\log 10 \text{ } CH^+$$

pH scale is used to measure the acidity or alkalinity of a soil solution (or other solution). This scale runs from 0 pH to 14 pH. In this scale, the 7 units level is known as neutral i.e. neither acidic nor alkaline. Pure water has a pH-7.0. All values below pH 7.0 denotes acidity and the values above pH 7.0 represents alkalinity. The degree of acidity increases as pH decreases below pH 7.0. Soil showing pH 5 is ten times more acidic than showing pH 6.0. Likewise, the degree of alkalinity increase as we go higher from pH 7.0. The alkalinity at pH 9.0. unit is ten times more than that pH 8.0 units.

Factors Controlling Soil Reaction

There are some factors that control soil reaction are as follows:

Nature of Soil Colloid

Soil colloid influences soil reaction to a very great extent. Soil colloids when dominated by adsorbed hydrogen (H^+) ion, the reaction of soil becomes acidic. On the other hand, soil colloid when dominated by hydroxyl (OH^-) ion, the reaction of soil becomes alkalinity.

Nature of Ion

The soil that contains more hydrogen ion than hydroxyl ions becomes acidic in reaction. When the aluminium ions are present in the soil, they react with water to liberate hydrogen ions, which increases the soil acidity.

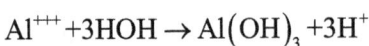

$$Al^{+++} + 3HOH \rightarrow Al(OH)_3 + 3H^+$$

Percentage Base Saturation

A low percentage base saturation of soil means soil acidity. In humid areas, the basic elements have been leached down from the soil, the percentage base saturation decreases much below 80 and they become acidic in reaction. If the percentage of base saturation is above 80 and at 90, then they become neutral in reaction and alkaline reaction respectively.

Rainfall

Rainfall plays important role in determining the soil reaction. The soils that are developed in high rainfall areas, becomes acidic in nature due to leaching of some nutrients such as calcium (Ca^{++}), magnesium (Mg^{++}) etc. from soil solution. So leaching encourages the development of soil acidity. On other hand, the soils that are developed in low rainfall areas, becomes alkaline in nature.

Fertilizers

The continual use of fertilizers is responsible for a marked change in soil pH. Acid forming fertilizers such as Ammonium sulphate, Urea, Ammonium nitrate etc. when applied in the soil in large quantities makes the soil acidic. On the other hand, basic fertilizers such as Sodium nitrate, Basic slag etc. makes the soil alkaline.

Influence of Soil Reaction on the Availability of Nutrients

Soil reaction affects the availability of nutrients to plants. Soil reaction affects indirectly in availability of nutrients to plants as soil organism do not function well in acid and alkali soil. Soil organism do their function at their best within a pH range 6.0-7.5.

The influence of soil reaction on the availability of nutrients to plant is as follows:

Nitrogen

Nitrogen is most important nutrient for plants. Plant absorbs nitrogen in the form of Ammonium (NH_4^+) and Nitrate (NO_3^-). Out of these two forms, plant absorbs most of their nitrogen in the form of nitrate (NO_3^-). The availability of nitrate nitrogen depends on the activity of nitrifying bacteria's. The microorganism responsible for nitrification are most active when the soil pH is between 6.5 and 7.5.

The activity of nitrifying bacteria is adversely affected if pH falls below 5.5 and more than 9.0. The activity of nitrogen fixing bacteria (e.g. Azotobacter) falls down at below soil pH 6.0. In acidic condition, the decomposition of organic matter, the main source of nitrogen, is also slow down.

Phosphorus

Phosphorus is an essential constituent of every living cells and for nutrition of plant and animal. The availability of phosphorus depends on the soil pH. In strongly acidic soil (pH 5.0 or less), iron, aluminium, magnesium and other bases remains in soluble form and phosphorus reacts with these bases are converted into insoluble form and become unavailable to plant.

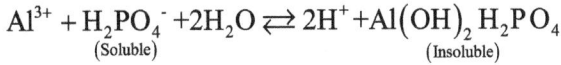

$$\underset{\text{(Soluble)}}{Al^{3+} + H_2PO_4^- + 2H_2O} \rightleftarrows \underset{\text{(Insoluble)}}{2H^+ + Al(OH)_2 H_2PO_4}$$

In acid soil, phosphorus become available to plants by anion exchange with the hydroxylanion (OH^-)

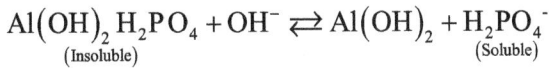

$$\underset{\text{(Insoluble)}}{Al(OH)_2 H_2PO_4 + OH^-} \rightleftarrows \underset{\text{(Soluble)}}{Al(OH)_2 + H_2PO_4^-}$$

One anion (OH^-) has been exchanged for another ion $(H_2PO_4^-)$. Phosphorus $(H_2PO_4^-)$ become available after liming in the acid soil.

Phosphorus fixation takes place even when the soil is alkaline. Phosphate ion combine with calcium ion and calcium or magnesium carbonate and form insoluble calcium or magnesium carbonate.

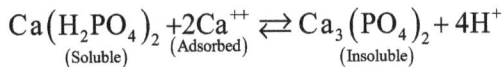

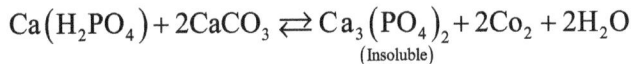

$$\underset{\text{(Soluble)}}{Ca(H_2PO_4)_2} + \underset{\text{(Adsorbed)}}{2Ca^{++}} \rightleftarrows \underset{\text{(Insoluble)}}{Ca_3(PO_4)_2} + 4H^+$$

$$Ca(H_2PO_4) + 2CaCO_3 \rightleftarrows \underset{\text{(Insoluble)}}{Ca_3(PO_4)_2} + 2Co_2 + 2H_2O$$

The availability of phosphorus remains highest at a soil pH between 6.5-7.5.

Potassium

Potassium is an essential element for the development of chlorophyll. The availability of potassium does not influence by soil reaction to any great extent. In acid soil, potassium is lost through leaching. Application of lime for reclamation of acid soil result in an increase in potassium fixation of soils and the potassium remains in the soil in the form of non-availability.

Sulphur

Sulphur is an important element for oil seeds, cruciferae, sugar and pulse crop. The availability of sulphur is not affected by soil reaction. In acid soil, it is more soluble and is subjected to loss by leaching.

Calcium

Calcium as calcium pectate is an important constituent of cell wall and requires in large amounts for cell division. Acid soils are poor in calcium. In alkali soil (pH not exceeding 8.5), the availability of calcium remains high. The availability of calcium decreases when soil pH is above 8.5.

Magnesium

Magnesium is an essential constituent of chlorophyll. Acid soils are poor in magnesium. In alkali soil (pH not exceeding 8.5), the availability of magnesium remains high. The availability of magnesium decreases when soil pH is above 8.5.

Manganese

Manganese is an essential constituent of chlorophyll and also formation of oils and fats. Soil pH has decided influence in the availability of manganese. At high pH values, all cations are unfavourably affected. Over liming or a naturally high pH is associated with deficiencies of manganese and such conditions occur in nature in many of the calcareous soils of West Bengal.

Iron

Iron is necessary for the synthesis of chlorophyll. In very acid soil, there is relative abundance of ions of iron. Iron deficiency of plant due to high pH is not uncommon. At high pH i.e. in alkali soils, ferrous (Fe^{2+}) ion is converted to ferric (Fe^{3+}) and precipitated as Ferric oxide (Fe_2O_3). The availability of iron increases as the pH of soil decreases.

Zinc

Soil pH affects the availability of Zinc. The zinc deficiency occurs on soils that are slightly acidic to neutral. High pH reduces the availability of zinc by precipitating zinc as zinc hydroxides.

Boron

Boron occurs in most soil in extremely small quantities. The availability and utilization of boron is determined to a considerable extent by soil pH. Boron is most soluble under acid condition. It apparently occurs in acid soil in part as boric acid and this is readily available to plants. The high soil pH causes boron deficiency in plants forming complex compound. A specific Ca : B is required for every crop. When calcium level is high, boron content should be high and if no, plant will show boron deficiency.

Copper

Copper is an essential constituent of enzyme. In very acid soil, there is relative abundance of copper. The solubility of copper decreases as pH increases particularly in sandy soils.

The decrease in solubility with increasing the soil pH may be result of precipitation of copper in the form of cupric oxide (CuO). The oxidized state of copper i.e. hydroxides or hydrous oxides is insoluble. Copper deficiency is induced by heavy liming and excessive application of nitrogen and phosphorus.

Molybdenum

Molybdenum is a constituent part of the enzyme, nitrate reductase. Molybdenum availability is significantly dependent on soil pH. It is quite unavailable in strongly acid soil and becomes available by liming of acid soil. As the pH is raised to 6.0 or above, its availability increases.

SOIL ACIDITY

Effects of Soil Acidity

Plant growth and most soil processes, including nutrient availability and microbial activity, are favoured by a soil pH range of 5.5 – 8. Acid soil, particularly in the subsurface, will also restrict root access to water and nutrients.

Aluminium Toxicity

When soil pH drops, aluminium becomes soluble. A small drop in pH can result in a large increase in soluble aluminium. In this form, aluminium retards root growth, restricting access to water and nutrients.

Poor crop and pasture growth, yield reduction and smaller grain size occur as a result of inadequate water and nutrition. The effects of aluminium toxicity on crops are usually most noticeable in seasons with a dry finish as plants have restricted access to stored subsoil water for grain filling.

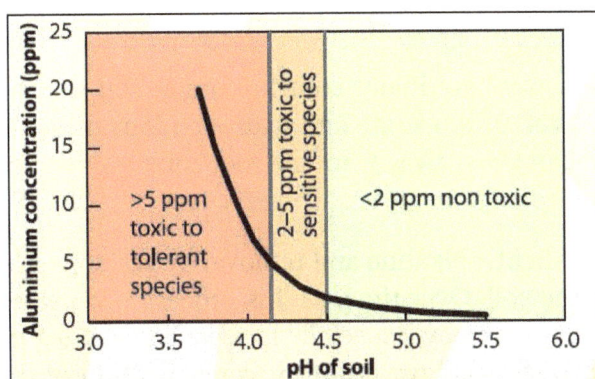

Al and pH graph with rule of thumb Al toxicity.

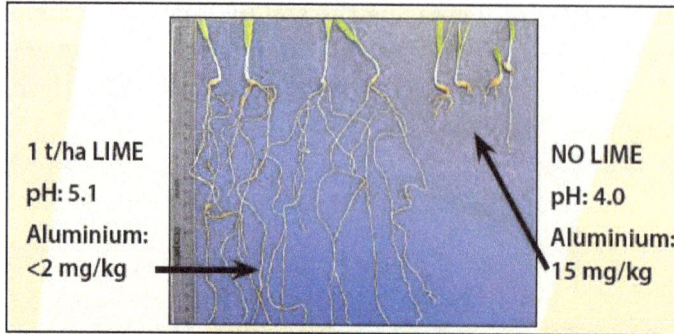

Roots of barley grown in acidic subsurface soil are shortened by aluminium toxicity.

Nutrient Availability

In very acid soils, all the major plant nutrients (nitrogen, phosphorus, potassium, sulphur, calcium, manganese and also the trace element molybdenum) may be unavailable, or only available in insufficient quantities. Plants can show deficiency symptoms despite adequate fertiliser application.

Microbial Activity

Low pH in topsoils may affect microbial activity, most notably decreasing legume nodulation. The resulting nitrogen deficiency may be indicated by reddening of stems and petioles on pasture legumes, or yellowing and death of oldest leaves on grain legumes. Rhizobia bacteria are greatly reduced in acid soils. Some pasture legumes may fail to persist due to the inability of reduced Rhizobia populations to successfully nodulate roots and form a functioning symbiosis.

Causes of Soil Acidity

Soil acidification is a natural process accelerated by agriculture. Soil acidifies because the concentration of hydrogen ions in the soil increases. The main cause of soil acidification is inefficient use of nitrogen, followed by the export of alkalinity in produce.

Ammonium based fertilisers are major contributors to soil acidification. Ammonium nitrogen is readily converted to nitrate and hydrogen ions in the soil. If nitrate is not taken-up by plants, it can leach away from the root zone leaving behind hydrogen ions thereby increasing soil acidity.

Most plant material is slightly alkaline and removal by grazing or harvest leaves residual hydrogen ions in the soil. Over time, as this process is repeated, the soil becomes acidic. Major contributors are hay, especially lucerne hay and legume crops. Alkalinity removed in animal products is low, however, concentration of dung in stock camps adds to the total alkalinity exported in animal production.

Management of Acidic Soils

Soil Testing

Knowledge of how soil pH profiles and acidification rates vary across the farm will assist effective soil acidity management.

Ideally, soil samples should be taken when soils are dry and have minimal biological activity. It is standard to measure pH using one part soil to five parts 0.01 M $CaCl_2$. Soils with low total salts show large seasonal variation in pH if it is measured in water. pH measured in water can read 0.6 – 1.2 pH units higher than in calcium chloride.

Soil sampling should take paddock variability into consideration. For example, clays have greater capacity to resist pH change (buffering) than loams, which are better buffered than sands. Samples should be taken at the surface and in the subsurface to determine a soil pH profile. This will detect subsurface acidity, which may underlie topsoils with an optimal pH.

Samples need to be properly located (e.g. GPS) to allow monitoring. Sampling should be repeated every 3 – 4 years to detect changes and allow adjustment of management practices.

Interpreting pH Results

Depending on soil pH test results, agricultural lime may need to be applied to maintain pH, or to recover pH to an appropriate level. If the topsoil pH is above 5.5 and the subsurface pH above 4.8, only maintenance levels of liming will be required to counter on-going acidification caused by productive agriculture.

If the topsoil pH is below 5.5, recovery liming is recommended. Keeping the topsoil above 5.5 will treat the on-going acidification due to farming and ensure sufficient alkalinity can move down and treat subsurface acidity.

Liming is necessary if the subsurface pH is below 4.8, whether or not the topsoil is acidic. If the 10 – 20 cm layer is below 4.8 but the 20 – 30 cm layer above 4.8, liming is still required. In this case the band of acidic soil will restrict root access to the more suitable soil below.

Liming

Liming is the most economical method of ameliorating soil acidity. The amount of lime required will depend on the soil pH profile, lime quality, soil type, farming system and rainfall. Limesand, from coastal dunes, crushed limestone and dolomitic limestone are the main sources of agricultural lime. Carbonate from calcium carbonate and magnesium carbonate is the component in all of these sources that neutralises acid in soil.

The key factors in lime quality are neutralising value and particle size. The neutralising value of the lime is expressed as a percentage of pure calcium carbonate which is given a value of 100 %. With a higher neutralising value, less lime can be used, or more area treated, for the same pH change. Lime with a higher proportion of small particles will react quicker to neutralise acid in the soil, which is beneficial when liming to recover acidic soil.

Complementary Management Strategies

If soil pH is low, using tolerant species/varieties of crops and pasture can reduce the impact of soil acidity. This is not a permanent solution because the soil will continue to acidify without liming treatment.

A number of management practices can reduce the rate of soil acidification. Management of nitrogen fertiliser input to reduce nitrate leaching is most important in high rainfall areas. Product export can be reduced by feeding hay back onto paddocks from where it has been cut. Less acidifying options in rotations will also help, e.g. replace legume hay with a less acidifying crop or pasture.

SOIL ALKALINITY

Soils with a pH level that is higher than 7 are said to be "alkaline." Such soils are suitable for growing plants that thrive in a "sweet" soil, as opposed to a "sour" or acid soil. If soil pH needs to be raised (that is, the ground is not alkaline enough), apply garden lime. If, on the other hand, your soil has too much alkalinity, you can lower the pH by applying a fertilizer that has sulfur/ammonium-N in it. When you are at the garden center, just look for a fertilizer intended for acid-loving plant.

Fortunately, just as there are plants that like acidic soils, which give you planting options on sour ground, so there are plants that like alkaline soil (or, at least, do not mind growing in it). Observe, however, that even plants within the same genus can "disagree" over what kind of ground that they like to grow in. Take the magnificent lady slipper orchid (Cypripedium), for example. There are many types. Some like their ground sweet, others like it sour, and still others prefer a soil pH that is somewhere in between. But generally speaking, the following types of plants are good choices to grow in alkaline soils.

Plants that Grow well in Alkaline Soils

The list below does not pretend to be exhaustive. But it gives you enough options to begin planning to landscape on ground that is alkaline. You will find perennials, vines, shrubs, and trees on the list. A plant's inclusion on this list does not necessarily mean that it needs or prefers to grow in an alkaline soil (although it might), only that it will, at the very least, tolerate alkalinity.

Perennials

Perennials (including mentions of a bulb plant and a couple of types of ornamental grasses), vines, shrubs, and trees. We will begin with perennials, which are the first plants that people think of when the subject of flower gardens comes up. By paying careful attention to sequence of bloom in your planning, you can experience many months of magnificent floral color in your yard by growing perennial flowers:

- Anchusa azurea
- 'Autumn Joy' sedum
- Bearded iris
- Black-eyed Susans
- Candytuft
- Catmint (Nepeta × faassenii)

Vines

Vines and particularly flowering vines are remarkably useful plants in a landscape. The one drawback they have, as a class (if one may generalize about them), is that many of them are aggressive. So if you are someone who, in your plant selection, strives to obtain plants that are compatible with low-maintenance landscaping, make it a point to research the qualities of a vine carefully before purchasing it:

- Boston ivy
- Clematis
- Kiwi (Actinidia kolomikta)
- Vinca minor
- Virginia creeper
- Winter jasmine

Shrubs (Bushes)

Shrubs have been dubbed the "backbone" of a landscape, because they furnish it will valuable structure. Select a variety of flowering shrubs if you wish to optimize color in your spring and summer landscape; some are also good shrubs for fall color. But also remember that the value of evergreen shrubs soars in winter, when all of your other plants have dropped their leaves. Some types of shrubs that will grow in alkaline soils are:

- Arborvitae

- Buxus sempervirens 'Suffruticosa'

- Contorted filbert

- Cotoneaster horizontalis

- Deutzia

Trees

Because of their size (and because of their corresponding cost and impact on your property), you have to pay special attention to the plant-selection process when choosing a tree. But if you get it right, you can end up with a plant that you will later deem indispensable to your landscape. Some are towering giants that can cast shade over a large portion of your yard, while others are much shorter and serve effectively as shade trees for patios. Still others are are considered dwarf trees and function more like shrubs in your landscaping:

- Common (or "European") beech

- European ash

- Ginkgo biloba

- Horse chestnut

- Mugo pine

- Ornamental cherry

References

- Soil-chemistry, earth-and-planetary-sciences: sciencedirect.com, Retrieved 17 May, 2019

- Queensland department of environment and heritage protection. "soil ph". Www.qld.gov.au. Retrieved 15 may 2017

- Measurement-of-ph-in-soil: bacto.com.au, Retrieved 19 April, 2019

- Sumner, malcolm e.; yamada, tsuioshi (november 2002). "farming with acidity". Communications in soil science and plant analysis. 33 (15–18): 2467–2496. Doi:10.1081/css-120014461

- Soil-reaction-types-factors-and-influence-soil-science, soil-reaction: soilmanagementindia.com, Retrieved 5 February, 2019

- Soil-acidity, factsheets: soilquality.org.au, Retrieved 26 July, 2019

- Soil-and-plants-that-dont-mind-alkalinity: thespruce.com, Retrieved 8 January, 2019

3

Soil Composition and Ionic Reactions

Soil has numerous chemical compounds present in it such as phosphoric acid, calcium and acid sulfate. One of the most important ionic reaction which takes place in the soil is the exchange reaction. This chapter closely examines these key components as well as ionic reactions to provide an extensive understanding of the subject.

COMPOSITION OF SOIL

Soil composition is an important aspect of nutrient management. While soil minerals and organic matter hold and store nutrients, soil water is what readily provides nutrients for plant uptake. Soil air, too, plays an integral role since many of the microorganisms that live in the soil need air to undergo the biological processes that release additional nutrients into the soil.

The basic components of soil are minerals, organic matter, water and air. The typical soil consists of approximately 45% mineral, 5% organic matter, 20-30% water, and 20-30% air. These percentages are only generalizations at best. In reality, the soil is very complex and dynamic. The composition of the soil can fluctuate on a daily basis, depending on numerous factors such as water supply, cultivation practices, and/or soil type.

The solid phase of soil, which includes minerals and organic matter, are generally stable in nature. Yet, if organic matter is not properly managed, it may be depleted from the soil. The liquid and gas phases of the soil, which are water and air respectively, are the most dynamic properties of the soil. The relative amounts of water and air in the soil are constantly changing as the soil wets or dries.

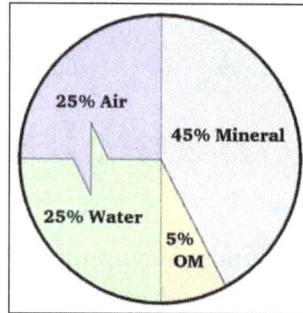

Approximate composition of soil.

Phosphoric Acid in Soil

All plants need the right balance of minerals and nutrients in order to thrive. Phosphate is one of those, and it is generally included in all fertilizers, whether organic in origin, or synthetic. Generally, this is done in the form of phosphoric acid, which is a derivative of phosphate.

Phosphoric acid cannot be taken up directly by plants – it must be changed first. After being applied to the soil, the acid is converted into hydrogen phosphate by chemical actions and biological agents. It is also transformed into dihydrogen phosphate, which is also usable by plants.

It should be noted that there are many primary sources of phosphorus and phosphoric acid. Animal manure contains phosphate, but it does not create new substances. It simply transports phosphate from one area to another (after the application of manure). However, by adding acid to animal bones, new phosphate can be created in the form of phosphoric acid.

The same process is used with phosphoric rock, which has become the primary source of phosphorus in commercial fertilizers. Acid is applied to the rock after it has been mined and crushed, thus rending phosphoric acid that can be applied in many different ways.

Note that phosphorous acid and phosphoric acids are very different. Phosphorous acid, which has one less oxygen molecule (H_3PO_3),Orthophosphoric acid; Trihydroxylphosphine oxide is used as a fungicide in diluted form. It is generally sold in salt form, but some liquid mixtures are available.

Calcium in Soil

Calcium has four basic functions in the soil which help to create better soil structure and healthy plants.

Calcium is an alkaline metal of Group II A on the periodic chart and is the fifth most abundant element in the earth's crust while being widely distributed in nature. The electronic structure of the calcium cation makes it a unique element that is ideally suited to assist in plant growth.

In the soil environment the calcium cation serves four basic functions that are critical to agricultural production.

In agricultural soils the calcium cation flocculates clay and organic matter particles which results in proper soil porosity. Proper soil porosity insures adequate soil aeration which guarantees proper soil drainage and allows correct root growth.

Soil biological life is the key to profitable plant nutrition and that biological life has a definite need for the element calcium. The beneficial biology of the soil complex are aerobic by nature and respond very favorably to the porosity of the soil complex which is provided by the flocculation of the exchange complex by the calcium cation. Also, soil biology feeds very heavily on available calcium in soil to supply the needs of their bodily functions.

All agricultural plants have a specific need for calcium in their tissue to insure proper growth. Many research studies have proven that animals consuming plants that have calcium deficiencies will encounter growth problems that may result in long term production losses.

The ability of calcium to neutralize either excess acid or excess alkaline, amphoterism, in the soil environment is the key to proper root growth of the plant. Without proper root growth plants are susceptible to drought and improper mineral nutrition.

The need for calcium in agricultural soil has been scientifically proven for many years. The proper levels of calcium need in soils has been questioned by those that benefit from its deficiency in soil. However, future basic nutritional research will demonstrate the lack of adequate calcium levels in modern agricultural soils.

Acid Sulfate Soil

Acid sulfate soils are naturally occurring soils, sediments or organic substrates (e.g. peat) that are formed under waterlogged conditions. These soils contain iron sulfide minerals (predominantly as the mineral pyrite) or their oxidation products. In an undisturbed state below the water table, acid sulfate soils are benign. However, if the soils are drained, excavated or exposed to air by a lowering of the water table, the sulfides react with oxygen to form sulfuric acid.

Release of this sulfuric acid from the soil can in turn release iron, aluminium, and other heavy metals (particularly arsenic) within the soil. Once mobilized in this way, the acid and metals can create a variety of adverse impacts: killing vegetation, seeping into and acidifying groundwater and surface water bodies, killing fish and other aquatic organisms, and degrading concrete and steel structures to the point of failure.

Formation

Polders with acid sulfate soils in Guinea Bissau along a sea-arm amidst mangroves.

The soils and sediments most prone to becoming acid sulfate soils formed within the last 10,000 years, after the last major sea level rise. When the sea level rose and inundated the land, sulfate in the seawater mixed with land sediments containing iron oxides and organic matter. Under these anaerobic conditions, lithotrophic bacteria such as *Desulfovibrio desulfuricans* obtain oxygen for respiration through the reduction of sulfate ions in sea or groundwater, producing hydrogen sulfide. This in turn reacts with dissolved ferrous iron, forming very fine grained and highly reactive framboid crystals of iron sulfides such as (pyrite). Up to a point, warmer temperatures are more favourable conditions for these bacteria, creating a greater potential for formation of iron sulfides. Tropical waterlogged environments, such as mangrove swamps or estuaries, may contain higher levels of pyrite than those formed in more temperate climates.

The pyrite is stable until exposed to air, at which point the pyrite rapidly oxidises and produces sulfuric acid. The impacts of acid sulfate soil leachate may persist over a long time, and/or peak seasonally (after dry periods with the first rains). In some areas of Australia, acid sulfate soils that drained 100 years ago are still releasing acid.

Chemical Reaction

When drained, pyrite- (FeS_2) containing soils (also called cat-clays) may become extremely acidic (pH < 4) due to the oxidation of pyrite into sulfuric acid (H_2SO_4). In its simplest form, this chemical reaction is as follows:

$$2FeS_2 + 9O_2 + 4H_2O \rightarrow 8H^+ + 4SO_4^{2-} + 2Fe(OH)_3 \downarrow$$

The product $Fe(OH)_3$, iron(III) hydroxide (orange), precipitates as a solid, insoluble mineral by which the alkalinity component is immobilized, while the acidity remains active in the sulfuric acid. The process of acidification is accompanied by the formation of high amounts of aluminium (Al^{3+}, released from clay minerals under influence of the acidity), which are harmful to vegetation. Other products of the chemical reaction are:

1. Hydrogen sulfide (H_2S), a smelly gas.

2. Sulfur (S), a yellow solid.

3. Iron(II) sulfide (FeS), a black/gray/blue solid.

4. Hematite (Fe_2O_3), a red solid.

5. Goethite ($FeO \cdot OH$), a brown mineral.

6. Schwertmannite a brown mineral.

7. Iron sulfate compounds (e.g. jarosite).

8. H-Clay (hydrogen clay, with a large fraction of adsorbed H^+ ions, a stable mineral, but poor in nutrients).

The iron can be present in bivalent and trivalent forms (Fe^{2+}, the ferrous ion, and Fe^{3+}, the ferric ion respectively). The ferrous form is soluble, whereas the ferric form is not. The more oxidized the soil becomes, the more the ferric forms dominate. Acid sulfate soils exhibit an array of colors ranging from black, brown, blue-gray, red, orange and yellow. The hydrogen clay can be improved by admitting sea water: the magnesium (Mg) and sodium (Na) in the sea water replaces the adsorbed hydrogen and other exchangeable acidic cations such as aluminium (Al). However this can create additional risks when the hydrogen ions and exchangeable metals are mobilised.

Geographical Distribution

Acid sulfate soils are widespread around coastal regions, and are also locally associated with freshwater wetlands and saline sulfate-rich groundwater in some agricultural areas. In Australia, coastal acid sulfate soils occupy an estimated 58,000 km², underlying coastal estuaries and floodplains near where the majority of the Australian population lives. Acid sulfate soil disturbance is often associated with dredging, excavation dewatering activities during canal, housing and marina developments. Droughts can also result in acid sulfate soil exposure and acidification.

Acid sulfate soils that have not been disturbed are called potential acid sulfate soils (PASS). Acid sulfate soils that have been disturbed are called actual acid sulfate soils (AASS).

Impact

Disturbing potential acid sulfate soils can have a destructive effect on plant and fish life, and on aquatic ecosystems. Flushing of acidic leachate to groundwater and surface waters can cause a number of impacts, including:

- Ecological damage to aquatic and riparian ecosystems through fish kills, increased fish disease outbreaks, dominance of acid-tolerant species, precipitation of iron, etc.

- Effects on estuarine fisheries and aquaculture projects (increased disease, loss of spawning area, etc.).

- Contamination of groundwater and surface water with arsenic, aluminium and other metals.

- Reduction in agricultural productivity through metal contamination of soils (predominantly by aluminium).

- Damage to infrastructure through the corrosion of concrete and steel pipes, bridges and other sub-surface assets.

Agricultural Impacts

Sea water is admitted to a bunded polder on acid sulfate soil for soil improvement and weed control, Guinea Bissau.

Potentially acid sulfate soils (also called cat-clays) are often not cultivated or, if they are, planted with rice, so that the soil can be kept wet preventing oxidation. Subsurface drainage of these soils is normally not advisable.

When cultivated, acid sulfate soils cannot be kept wet continuously because of climatic dry spells and shortages of irrigation water, surface drainage may help to remove the acidic and toxic chemicals (formed in the dry spells) during rainy periods. In the long run surface drainage can help to reclaim acid sulfate soils. The indigenous population of Guinea Bissau has thus managed to develop the soils, but it has taken them many years of careful management and toil.

Also in the Sunderbans, West Bengal, India, acid sulfate soils have been taken in agricultural use.

A study in South Kalimantan, Indonesia, in a perhumid climate, has shown that the acid sulfate soils with a widely spaced subsurface drainage system have yielded promising results for the cultivation of upland rice, peanut and soybean. The local population, of old, had already settled in this area and were able to produce a variety of crops (including tree fruits), using hand-dug drains running from the river into the land until reaching the back swamps. The crop yields were modest, but provided enough income to make a decent living. Reclaimed acid sulfate soils have a well-developed soil structure; they are well permeable, but infertile due to the leaching that has occurred.

In the second half of the 20th century, in many parts of the world, waterlogged and potentially acid sulfate soils have been drained aggressively to make them productive for agriculture. The results were disastrous. The soils are unproductive, the lands look barren and the water is very clear, devoid of silt and life. The soils can be colorful, though.

Construction

When brickwork is persistently wet, as in foundations, retaining walls, parapets and chimneys, sulfates in bricks and mortar may in time crystallise and expand and cause mortar and renderings to disintegrate. To minimise this effect specialised brickwork with low sulfate levels should be used. Acid sulfates that are located within the subsoil strata has the same effects on the foundations of a building. Adequate protection can exist using a polythene sheeting to encase the foundations or using a sulfate-resistant Portland cement. To identify the pH level of the ground a soil investigation must take place.

Restoration and Management

By raising the water table, after damage has been inflicted due to over-intensive drainage, the soils can be restored. The following table gives an example.

Drainage and yield of Malaysian oil palm on acid sulfate soils Yield in tons of fresh fruit per ha:

Year	60	61	62	63	64	65	66	67	68	69	70	71
Yield	17	14	15	12	8	2	4	8	14	19	18	19

Drainage depth and intensity were increased in 1962. The water table was raised again in 1966 to counter negative effects.

In the "millennium drought" in the Murray-Darling Basin in Australia, exposure of acid sulfate soils occurred. Large scale engineering interventions were undertaken to prevent further acidification, including construction of a bund and pumping of water to prevent exposure and acidification of Lake Albert. Management of acidification in the Lower Lakes was also undertaken using aerial limestone dosing.

EXCHANGE REACTIONS IN SOIL

Soil colloids exhibit charges at the mineral surface, which arise from structural defects and protonation reactions. Since charge balance must be maintained counter ions of opposite charged satisfy most of the surface charges. The interaction varies from covalent bonding (specific adsorption) to weak electrostatic interactions. Ions attracted by weak electrostatic interactions can be replaced by other similarly charge ions. This process of stoichiometric replacement has been termed exchange. Interaction of positive ions is called cation exchange and conversely, negatively charged ions participate in anion exchange.

Exchange reactions are rapid, if the site is solution accessible, stoichiometric, reversible, and exhibits preferences (selectivity) among ions of differing charge and size.

Since exchange is a chemical reaction, it can be treated like any other mass action expression and an equilibrium constant can be defined for the reaction. The equilibrium expression K has been named the selectivity coefficient, exchange coefficient and other terms.

For the reaction: $AX + B \rightleftarrows BX + A$

We may define the following selectivity coefficient:

$$\text{Selectivity coefficient} = \ K_B^A = \frac{(BX)(A)}{(AX)(B)}$$

If the value of the selectivity coefficient is >1, this implies that B is preferred over A and if K <1 then A is preferred over B. There are several reasons that one ion may be more strongly attracted to the surface than another ion.

Ion Exchange Selectivity

The following is a discussion of some of the reasons for differential preference:

- Valence - The general rule is that the exchanger prefers the ion of higher valence (more properly, the charge on the ion). However, we will see later that under some conditions, this rule can be violate.

- Ion Solvation - The exchanger will prefer the ion that is less solvated with water molecules. This rule also takes the form of the "Iyotropic series". Another formulation is that the exchanger will prefer an ion which is a (water) structure "breaker" over one which is a structure "maker".

- Chelation By Exchanger - Some exchangers such as proteins, humates, fulvates and polyuronides have vicinal functional groups that can form a multidentate complex with ions. Since divalent (and higher valent) ions generally form chelate structures whereas monovalent ions do not, the above environmental materials are likely to show additional preference for higher valent ions.

- Van der Waal's Forces - The exchanger will prefer the ion with the greater number of "exposed" atoms which can form dipole-induced dipole attractive forces between the exchanger and the ion. An example of this is the inability of mineral salts to displace polynuclear aluminum hydroxides from the interlayer of smectites and vermiculites.

- Sieve Action by Exchanger - Zeolites are aluminosilicate exchangers with internal channels of sizes that permit small ions to enter but which exclude larger ions.

Exchange Process

A number of expressions have been proposed to quantitatively describe the equilibrium between ions in bulk solution (pore water) and ions in aqueous solution immediately adjacent to charged colloid surface (interface). The first type is referred to as "solution phase" ions that are no different in concept from ions that have come into solution from dissolution of salts while the second type is called "exchangeable" ions. Some of these types of ions are thought to be loosely bound to the surface by forces in addition to electrostatic forces while others are clearly dissociated from the surface and fully hydrated but still held in the area by electrostatic forces. The boundary between the exchange "phase" in solution and the "solution" is not sharp or distinct; hence the term diffuse layer to describe the zone in solution which includes all of the exchange ions. Although there is a distressing vagueness to this general picture, we use a pragmatic experimental approach which says that solution phase composition for a column of soil or sediment in a cylinder is the salt composition of an aqueous solution that passes into the top of the column and exits the bottom of the column without change. By definition, the exchange phase is at equilibrium with a solution of exactly known ionic composition. Next, we extract all of the pore water (including its salt) as well as all the exchangeable ions. By subtracting the quantity of salt ions from the total quantity of ions, the quantity of exchange ions in the sample is estimated. Obviously, this simple approach assumes there is one type of charge in the sample, cation exchange capacity or anion exchange capacity but not both.

Solution phase ion activities are expressed as usual:

$$\left(Ca^{2+}\right) - \text{Calcium ion activity} = ?_{Ca}\left(Ca^{2+}\right).$$

where $?_{Ca}$ = the activity coefficient of calcium ion in that particular solution and $\left(Ca^{2+}\right)$ = the concentration of free calcium ion in solution. Many different nomenclature conventions are used throughout the literature to represent the exchange phase composition in equilibrium expressions:

$$\overline{Ca}, \quad CaX, \quad Ca_X, \quad Ca\text{-Clay}$$

In addition to these obvious but unimportant differences, one must be careful to note that different expressions have quite different dimensions. For instance, the selectivity coefficient of Helfferich (K_B^A), includes terms for A and B in dimensions of moles A (and B) per liter of exchange phase fluid. It is fairly easy to estimate the volume of water inside a bead of swollen synthetic organic resin exchanger, and since most of the exchange capacity (the sum of the functional groups) of the bead is inside the bead, it is possible in this special exchanger to express AX (or BX) in the dimensions given above. Obviously, it is very difficult unless one makes some very arbitrary assumptions to estimate (in a soil-water system) an exchange phase fluid volume. When estimated, this volume is sometimes termed an "exclusion volume", again because salt is virtually excluded from this zone of water in the pores. Note in the selectivity coefficient expression, values of AX and [A] are both given in moles per liter but the exchange solution is a different compartment than the salt solution compartment (pore water). In making measurements on real columns of resin bead exchangers, the total volume of water inside all the beads in the column, V_{BT} is likely to be different from the total volume of all the water in the pores, V_{PT} between the beads. In order to calculate [A], one uses a chemical estimate of the total quantity of salt A, Q_A, and V_{PT}:

$$[A] = Q_A / V_{PT}.$$

Note that later in the simplified dynamic exchanger model, Farmer Fillipi, we make the assumption that $V_{PT} = V_{BT}$.

This difficulty (knowing V_{BT}) is eliminated by the Vanselow convention which uses the mole fractions of the ions in the exchanger, N_A and N, as the representing dimensions:

$$N_A = \frac{Q_A}{Q_A + Q_B}$$

This does introduce another minor difficulty when the valences of the two ions, z_A and z_B are not equal because the total exchange capacity in the column, CEC, in dimensions of equivalents is:

$$CEC = z_A Q_A + z_B Q_B$$

Therefore, when $z_A \neq z_B$, $Q_A + Q_B \neq$ constant at different Q_A/Q_B, even when CEC is constant.

Following are statements and definitions about some of the exchange expressions that have been proposed. Their relative quality has been or can be Judged on two bases. Some expressions are preferred by chemists because they have a solid basis in thermodynamics, i.e., they can be derived by formal chemical thermodynamic arguments very similar to those invoked to derive expressions for acid-base dissociation, ion-pair formation, chelation, complex formation, or solubility "products". (It is not expected at the level of this class that the student has knowledge of chemical thermodynamic theory, but it is expected that you have already had some experience with the working expressions derived from thermodynamics for these equilibria.) A second criteria used to judge quality of an expression is the degree to which the expression describes equilibria over a wide range of chemical compositional space, i.e. from $Q_A/(Q_A + Q_B) \sim 0$ to $Q_A(Q_A + Q_B) \sim 1$.

None of the expressions given below are really good when judged on both criteria simultaneously. From a practical standpoint, the second criterion is certainly the most important.

Thermodynamic Expressions

A general stoichiometric reaction is:

$$aA + b\overline{B} = a\overline{A} + bB$$

This does not give the valences of A and B but the integer values of (a) and (b) are based on the integer values of these valences. The main reason we write the above expression is to see what the values of (a) and (b) should be and it is not important whether the expression is written as above or from right to left with $\overline{A}$ and B as "reactants" rather than as "products".

The equilibrium thermodynamic expression is:

$$K = \frac{\left(\overline{A}\right)^a (B)^b}{\left(\overline{B}\right)^b (A)^a}$$

where ( ) = activity of the ion in the phase.

A specific example of monovalent-divalent cation exchange in soil represented by the symbol X is:

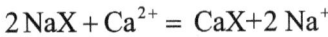

$$2\,NaX + Ca^{2+} = CaX + 2\,Na^+$$

$$K = \frac{(CaX)\left(Na^+\right)^2}{(NaX)^2\left(Ca^{2+}\right)}$$

Note that the thermodynamic K has no subscripts or superscripts. Also note that the activity of the participant ions can be calculated with standard techniques using total concentrations of the free ions and the ionic strength of the solution. However, there are no generally accepted techniques to calculate the activity coefficients of ions in the exchanger phase, $\overline{\gamma}$, so that for example:

$$(\overline{A}) \quad \overline{\gamma}[\overline{A}]$$

One of the first suggestions to deal with this problem was that of Vanselow who suggested that mole fraction, N_i, of an ion in the exchange phase might be proportional to activity in the exchange phase by analogy with the known behavior of two different elements in solid solutions such as copper and zinc in brass. Thus,

$$(\overline{A}) = \overline{\gamma_A} N_A$$

where,

$$N_A = \frac{Q_A}{Q_A + Q_B}$$

Without calculating the $\overline{\gamma}$'s, the Vanselow exchange coefficient, Kv, is given by:

$$Kv = \frac{N_A^a\,(B)^b}{N_B^b\,(A)^a}$$

Therefore, $K = \dfrac{\overline{\gamma}_A^{-a}}{\overline{\gamma}_B^{-b}}\,Kv$

For the specific case of Na-Ca exchange:

$$Kv = \frac{N_{ca}\left(Na^+\right)^2}{N^2_{Na}\left(Ca^{2+}\right)}$$

Initially, it was probably hoped that Kv would be relatively constant at different values

of the ratio $\dfrac{(B)^b}{(A)^a}$. This turned out to be true but only over narrow ranges of the ratio.

Therefore, one would expect this to be the death of the Vanselow expression. It was not, for two reasons. First, a mole fraction is an easy parameter to calculate from experimental data because it does not involve the additional uncertainty of the value of VPT in a given system. Second, Gaines and Thomas developed a theory based on further thermodynamic arguments that lead to the estimation of K from many measurements

of K_V at many selected experimental values of $\dfrac{(B)^b}{(A)^a}$. We will not study this theory be-

cause even though we may know K for a particular sample of a soil or a sediment, it is not possible to use K to predict the exchange behavior of that material in the field without also knowing each of the exchange phase activity coefficients as functions of exchange phase composition. The cost of doing this for a soil profile by standard techniques would be exorbitant. Sparks details the methods of deriving K from K_v because the theoretical applications of exchange equations are problematic, less rigorous formulations have been developed to describe cation exchange in soils.

Empirical or Semi-theoretical Expressions

Gapon

$$K_G = \frac{[Ca_{1/2}X]\left[Na^+\right]}{[NaX]\left[Ca^{2+}\right]^{0.5}}$$

As ordinarily applied, the cation concentrations in solution are inserted in dimensions of mmol/L (concentration, not activity), while the exchange composition is inserted in dimensions of meq/100 g of sample. Note that the expression includes both exchangeable sodium and exchangeable calcium with a unitary exponent. This equation was adopted by workers at the USDA Salinity Laboratory at Riverside, California to describe the exchange behavior of Western United States soils. They then collected many soils with a variety of salinity and Na-Ca-Mg compositions. Soluble and exchange ion compositions of each were measured. Then, the data were plotted with NaX/(CaX + MgX) on the Y-axis and $(Na^+)/(Ca^{2+} + Mg^{2+})^{0.5}$ on the X-axis. The parameter on the Y-axis was dubbed Exchangeable Sodium Ratio (ESR) while that on the X-axis was called the Sodium Adsorption Ratio (SAR). The "best" straight line was fitted through the data graphically giving the equation:

$$ESR = -0.01 + 0.015 \, (SAR)$$

This expression is sometimes called the statistical regression equation. Note that the slope $\sim 1/K_G$ and that this form is tantamount to assuming that:

$$K = 1 = \frac{CaX \left[Mg^{2+} \right]}{MgX \left[Ca^{2+} \right]}$$

The equation applies reasonably well to saline or saline-alkali soils dominated by smectite clay minerals, but should not be applied without calibration to variable charge soils.

There are a number of other exchange equations in the literature that have been used to describe cation exchange in soils. However, the discussion above will suffice to introduce you to the concept of exchange equations and selectivity. Application of exchange equations to soil systems produce some interesting results in relation of changes in soil moisture, and salt levels.

Dilution (or Alternatively, Desiccation)

Lets calculate the effect of changing salt level (dilution or desiccation) on composition of the exchange and solution phase.

Assume:

1. $K_V = 1 = constant = \dfrac{N_{Ca} \left[Na^+ \right]^2}{N_{Na}^2 \left[Ca^{2+} \right]}$.

2. ESP = % of exchange phase occupied by Na^+.

   ECaP = % of exchange phase occupied by Ca^{2+}.

3. All = 1.

Solution phase	Exchange phase	
$(Ca^{2+})/(Na^+)$	ESP	EcaP
.001/001	1.58	98.4
.01/.01	4.99	95.0
0.1/0.1	15.6	84.4
1.0/1.0	44.7	55.3
10/10	84.5	15.5

The result is that exchangeable sodium percentage is not constant. The exchanges shows increasing preference for the lower valent ion as the concentration of both ions increase. The phenomenon has been formalized by Schofield as the "Ratio Law".

Exchange Composition during Desiccation of Solution Phase

The following graphs were constructed from calculations resulting from the following models, assumptions and protocol:

Initially the soil is saturated at 0.385 L H_2O/kg soil. This solution is considered to have two different initial compositions (m_i):

a.   Case 1. $\left[Na^+\right] = \left[Ca^{2+}\right] = 0.005$ moles/L $= m_i$

b.   Case 2. $\left[Na^+\right] = \left[Ca^{2+}\right] = 0.015$ moles/L $= m_i$

The exchange equation used is that due to Vanselow arranged in the following form:

$$\frac{\left[Ca_X + Na_X\right]\left(Na^+\right)^2}{\left[Na_X\right]^2 \left(Ca^{2+}\right)} = Kv$$

where:

(Na^+) = moles Na^+ /L solution phase

(Ca^{2+}) = moles Ca^{2+}/L solutlon phase

$[Ca_X]$ = moles Ca^{2+} in the exchange phase / kg O.D. soil

$[Na_X]$ = moles Na^+ in the exchange phase / kg O.D. soil

Two cases are considered:

- $K_V = 1$

- $K_V = 5$

Exchange capacity of the soil is constant at 150 mmol (+)/kg O.D. soil.

After the first computation is performed at 0.385, the soil is dried without gain or loss of electrolyte from the system.

One special case is considered where $K_V = 1$, $m_i = 0.015$, but where the concentration of Ca^{2+} in solution remains constant during drying due to the precipitation of gypsum.

Two parameters are calculated during the drying process which are defined as:

a. f_{Na} in solution = equivalent fraction of Na^+ in solution:

$$f_{Na} = \frac{\left(Na^+\right)}{2\left(Ca^{2+}\right)+\left(Na^+\right)}$$

b. f_{Nax} - equivalent fraction of Na in exchange phase:

$$f_{NaX} = \frac{\left(Na_X\right)}{2\left(Ca_X\right)+\left(Na_X\right)}$$

Calculated changes in exchanger and solution phases during in situ drying:

- $K_V = 5$

- $Ca_i^{2+} = Na_i^+ = 0.005$

		Na^+	Ca^{2+}	f_{Na^+}	f_{NaX}
0.385		.00500	.00500	.333	.0158
0.340		.00549	.00575	.323	.0162
0.300		.00603	.00661	.313	.0166
0.250		.00690	.00810	.299	.0171
0.200		.00812	.01040	.281	.0178
0.150		.00997	.01430	.259	.0187
0.100		.01320	.02230	.229	.0198
0.050		.02100	.04720	.182	.0216

- $K_V = 1$

- $Ca_i^{2+} = Na_i^+ = 0.005$

Na^+	Ca^{2+}	f_{Na^+}	f_{NaX}
.00500	.00500	.333	.0353
.00545	.00577	.321	.0358
.00593	.00666	.308	.0363
.00671	.00820	.290	.0370
.00778	.01050	.269	.0378
.00940	.01460	.244	.0388
.01210	.02280	.210	.0401
.01850	.04850	.160	.0420

- $K_V = 1$

- $Ca^{2+} = 0.015 = constant,\ Na_i^+ = 0.015$

	Na^+	Ca^{2+}	f_{Na^+}	f_{NaX}
0.385	.0150	.015	.333	.0611
0.340	.0157	.015	.344	.0640
0.300	.0164	.015	.354	.0668
0.250	.0174	.015	.367	.0707
0.200	.0184	.015	.381	.0750
0.150	.0197	.015	.396	.0800
0.100	.0210	.015	.412	.0856
0.050	.0227	.015	.43	.092
0.020	.0237	.015	.442	.0965

- $K_V = 1$

- $Ca_i^{2+} = Na_i = 0.015$

Na^+	Ca^{2+}	f_{Na^+}	f_{NaX}
.0150	.0150	.333	.0611
.0164	.0173	.322	.0624
.0180	.0199	.311	.0636
.0205	.0244	.296	.0655
.0240	.0313	.277	.0676
.0293	.0431	.254	.0703
.0385	.0674	.222	.0740
.0604	.1430	.174	.0795
.106	.3800	.122	.0855

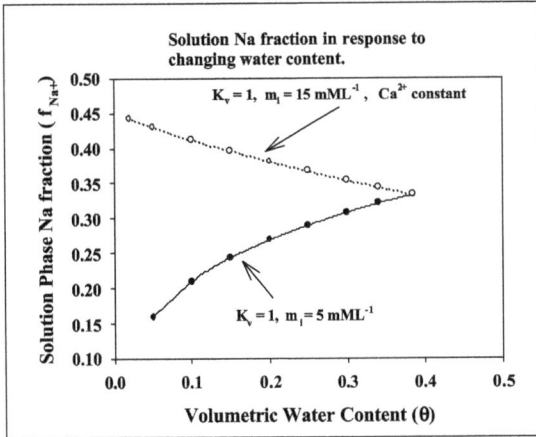

Solution Na fraction in response to changing water content.

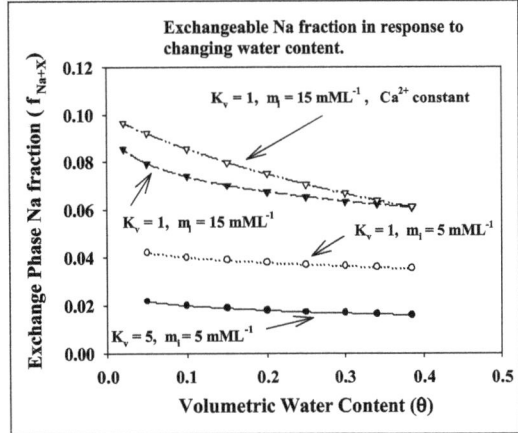

Exchangeable Na fraction in response to changing water content.

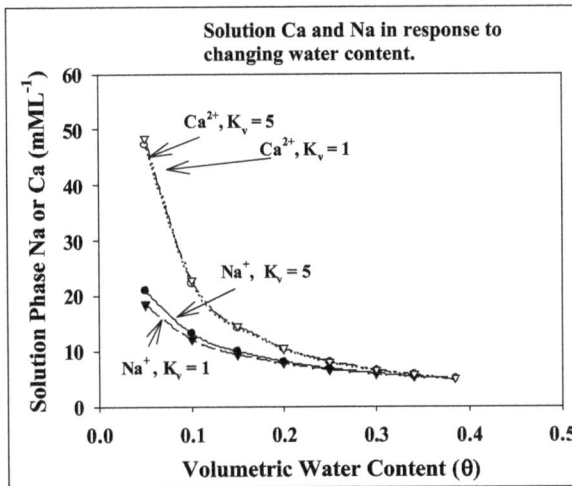

Solution Ca and Na in response to changing water content.

Beckett - Q/I Relationships

Exchange reaction and exchange capacity in soils are sources of plant nutrient such as K, Ca and Mg. However, it is difficult to visualize how the complex interaction between ions, exchange capacity and extraction of nutrients by plant affect nutrient supplies. Beckett developed a technique and theory to examine the buffer capacity of soils with respect to potassium. The theory is based on exchange reactions between Ca, Mg, and K. Beckett called the buffering capacity of the soil with respect to potassium quantity-intensity relations. In this concept Q stands for quantity, I stands for intensity. Although the following from Beckett was derived for assessing potassium status in soils, it can be generalized to a large number of other systems. These are summarized and presented in the following table.

System	Q	I	Buffer Index (dQ/dI)
Acid-base	C_a or C_b	Ph	dC/dpH
thermal	calories	temperature	Dq/dT

Soil water	% water in soil	Vapour pressure	d? /dp $_{H2O}$
Soil nutrient-K⁺	$\overline{K}$	AR^K	$\dfrac{d\,\overline{K}}{d\,AR^K}$

The following graph illustrates the concept of Q/I for soil solution ions and soil buffering capacity as derived from the changes in Q in relation to changes in solution ion levels.

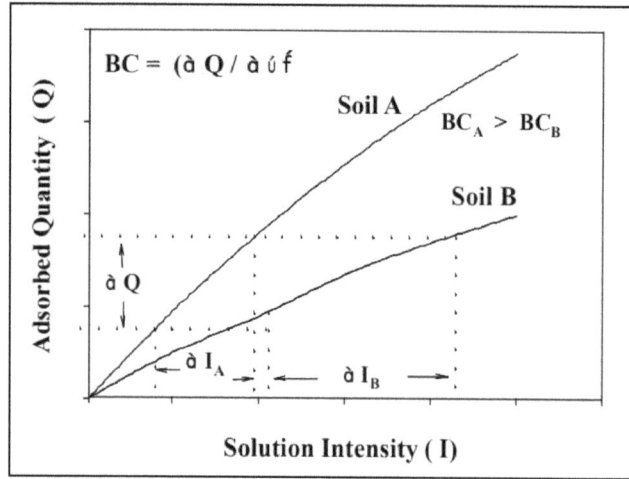

Beckett's proposal for evaluation of soil potassium status is based on a mass action exchange equation:

$$K_{K^+}^{M^{2+}} = \frac{[\overline{Ca^{2+} + Mg^{2+}}](K^+)^2}{[K^+]^2(Ca^{2+}+Mg^{2+})}$$

where ( ) = activities, and [ ] = concentrations. Take the square root and rearrange:

$$\overline{[K^+]} = \frac{(K^+)[\overline{Ca^{2+} + Mg^{2+}}]^{0.5}}{(Ca^{2+}+Mg^{2+})^{0.5}(K_{K^+}^{M^{2+}})^{0.5}}$$

In the above, the Q or quantity term for potassium = $\overline{[K^+]}$ while $\left(\dfrac{(K^+)}{(Ca^{2+} + Mg^{2+})^{0.5}}\right) =$ AR^K the activity ratio in solution is the I or intensity term for potassium.

If Q is plotted as a function of I, the slope should represent the ratio of exchangeable calcium plus exchangeable magnesium divided by the square root of the selectivity coefficient.

However, Beckett added an additional feature permitting the examination of the change in exchangeable potassium and obviating the need to directly measure $\overline{[K^+]}$.

Suppose that AR^K is changed so that AR^K increases. The equation predicts an increase in $[\overline{K^+}] = \Delta \overline{K^+}$. This new condition could be represented as follows:

$$[\overline{K^+}] = \Delta [\overline{K^+}] = \frac{(K^+)\left(\overline{Ca^{2+}} + \overline{Mg^{2+}} - \dfrac{\overline{\Delta K^+}}{2}\right)^{0.5}}{\left(Ca^{2+} + Mg^{2+}\right)^{0.5}\left(K_{K^+}^{M^{2+}}\right)^{0.5}}$$

Rearranging:

$$\Delta [\overline{K^+}] = \frac{(K^+)\left(\overline{Ca^{2+}} + \overline{Mg^{2+}} - \dfrac{\overline{\Delta K^+}}{2}\right)^{0.5}}{\left(Ca^{2+} + Mg^{2+}\right)^{0.5}\left(K_{K^+}^{M^{2+}}\right)^{0.5}} - [\overline{K^+}]$$

If $\dfrac{[\overline{\Delta K^+}]}{2} << [\overline{Ca^{2+}} + \overline{Mg^{2+}}]$ then the slope of a plot of $[\overline{K^+}]$ versus AR^K should be approximately constant and this function should approximate a straight line with a negative intercept equal to exchangeable potassium in the original sample.

The procedure is to weigh out several 5 g soil samples. To each is added a KCl and CaCl$_2$ solution of designed potassium and calcium concentration. Fifty mL of solution is added to each soil sample and the flask is shaken. Then it is centrifuged or filtered and K^+, Ca^{2+} and Mg^{2+} are measured. Ionic strength (μ) is calculated and activity coefficients for the three cations are calculated from by the extended Debye-Hückel (or other) equation. These values are used to calculate AR^K. Differences between the final equilibrium solution K^+ and the initial solution K^+ are $\dfrac{\Delta(\Delta K)}{\Delta AR^K}$ due to adsorption or desorption of exchangeable K. The x axis value is a measure of available K+ or the intensity of labile K in a soil. The potential buffering capacity (PBC^K) is a measure of the ability of the soil to maintain the intensity of K in soil solutions. It is proportional to CEC. Low PBC^K indicates a need for frequent fertilization. K_X is a measure of specific sites for K, and $K°$ is the labile or exchangeable K.

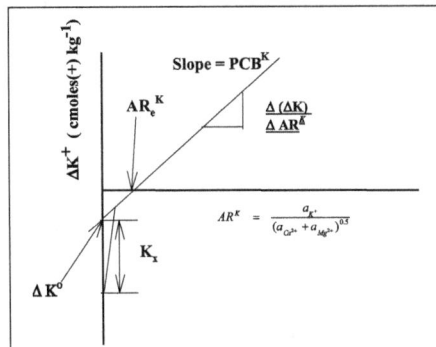

A typical Quantity/Intensity (Q/I) plot from Sparks and Liebhardt 1981. Effects of Long-term Lime and Potassium Applications on Quantity-Intensity Relationships in Sandy Soil.

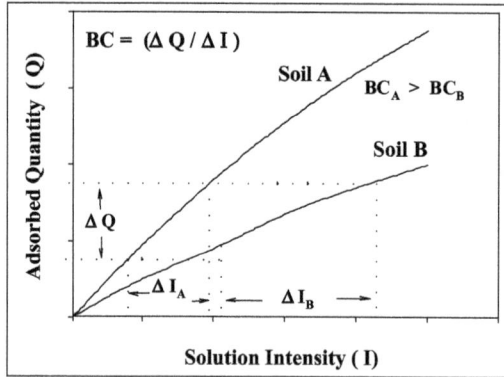

"Beckett" plot of potassium equilibrium data from two soils.

Cation Exchange by Cartoon Strip

The following presentation describes the interaction of a soil via exchange with an irrigation water containing only Ca^{2+} and Mg^{2+} in a hypothetical cotton field owned by Farmer Filippi. The physical picture of this interaction is to treat the soil a number of small soil volumes which are very thin. (In essence this is the formulation similar to plates in a condenser or in a chromatographic column.) The figure below depicts the system under discussion. Each frame for the cartoon strip represents a small increment in time as the initial solution infiltrates downward in the soil. The solutions are not allowed to mix and therefore piston displacement is assumed for all of the water in the system. A selectivity of 3 for Ca is assumed along with an irrigation water composition of 50 Ca and 25 Mg.

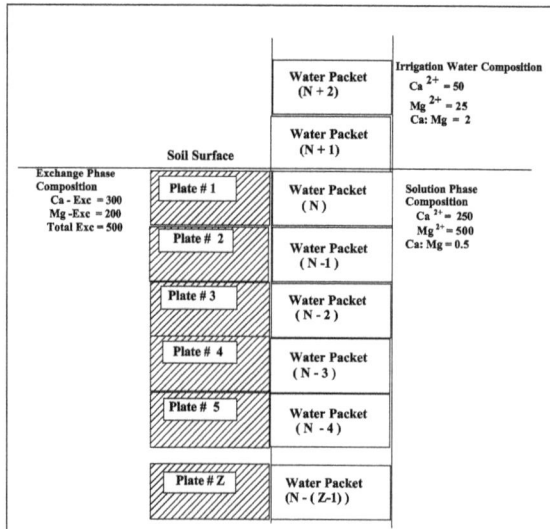

A schematic representation of irrigation water interacting with thin plates of soil.

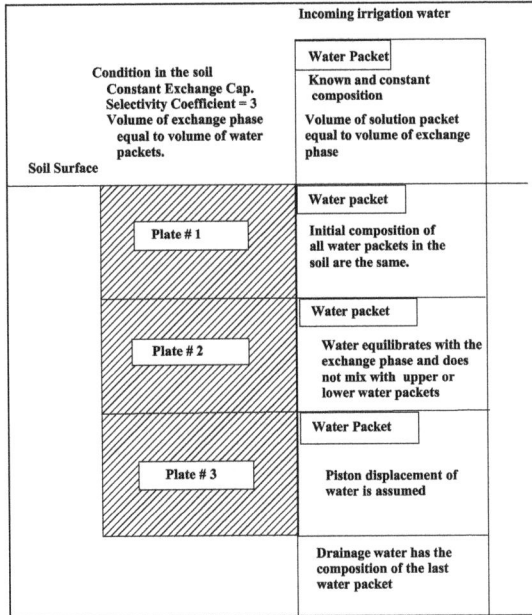

Conditions in the water packets and soil solid phase as irrigation
water of known composition infiltrates into a soil pore.

Plate 1	Soil Solution
Ca-Ex=300	Ca=250
Mg-Ex=200	Mg=500
Plate 2	Soil Solution
Ca-Ex=300	Ca=250
Mg-Ex=200	Mg=500

Exchanger phase (left) is at equilibrium with the solution phase (right) in the surface
of a field about to be irrigated with with water containing 50 Ca and 25 Mg. $K^{Ca}_{Mg} = 3$
= (CaX)(Mg) / (M$_g$X) (Ca).

	Water Packet 2
	Ca=50
	Mg=25
Plate 1	Soil Solution (WP 1)
Ca-Ex=300	Ca=250
Mg-Ex=200	Mg=500

Farmer Fillipi applies water to the field. Note that this water is more dilute than the soil
water and that Ca/Mg in the irrigation water is 2.0 while in the soil water it is 0.5. The

incoming water is called water packet 2 (WP2). The thin layer of soil is called plate 1 (PL 1). A plate is of the order of 3-4 mm thick.

	Water Packet 3 Ca=250 Mg=500
Plate 1 Ca-Ex=300 Mg-Ex=200	Water Packet 2 Ca=50 Mg=25
	Soil Solution (WP 1) Ca = 250 Mg = 500

The packet of water displaces an equal volume of soil water without mixing. No exchange or reaction has occurred (yet) between the incoming water and the exchangeable ion on the soil. The equilibrium values of the exchangeable and solution ions can be calculated knowing the selectivity coefficient and including a mass balance statement.

Plate 1 Ca-Ex = 321.8 Mg-Ex = 178.2	Water Packet 2 Ca = 28.2 Mg = 46.8
Plate 2 Ca-Ex = 300 Mg-Ex = 200	Soil Solution (WP 1) Ca = 250 Mg = 500

The system comes to equilibrium. Values in Frame must be used to solve equation of the Ca-Mg derivation sheet to give these equilibrium values.

	Water Packet 3 Ca = 50 Mg = 25
Plate 1 Ca-Ex = 321.8 Mg-Ex = 178.2	Water Packet 2 Ca = 28..2 Mg = 46.8

A new water packet (WP 3) is position to infiltrate into the soil displacing WP2 to the next soil plate PL 2.

Plate 1	Water Packet 3
Ca-Ex = 321.8	Ca = 50
Mg-Ex = 178.2	Mg = 25
Plate 2	Water Packet 2
Ca-Ex = 300	Ca = 28.2
Mg-Ex = 200	Mg = 46.8
	Soil Solution (WP 1)
	Ca = 250
	Mg = 500

The water packet (WP3) has entered and displaced the previous equilibrated packet. The displace packet is now opposite another volume of exchanger having the same composition as Plate 1 before infiltration of water packet 2 (WP2). The system is not at equilibrium.

Plate 1	Water Packet 3
Ca-Ex = 340.6	Ca = 31.2
Mg-Ex = 159.4	Mg = 43.8
Plate 2	Water packet 2
Ca-Ex = 302.8	Ca = 25.39
Mg-Ex = 197.2	Mg = 49.6

Both layers come to equilibrium.

Plate 1	Water Packet 4
Ca-Ex = 356.6	Ca = 34
Mg-Ex = 143.4	Mg = 41
Plate 2	Water Packet 3
Ca-Ex = 321.8	Ca = 26.1
Mg-Ex = 178.2	Mg = 48.9
Plate 3	Water Packet 2
Ca-Ex = 340.6	Ca = 25.04
Mg-Ex = 159.4	Mg = 49.95
Plate 4	Water Packet 1
	Ca =
	Mg =

A third packet enters and all layer have come to a new equilibrium. What is the composition of plate 4 and water packet 1 in Frame?

Derivation of the Ca-Mg Exchange Function

A. Selectivity coefficient:

$$K^{Ca}_{Mg} = \frac{[\overline{Ca}]\,[Mg^{2+}]}{[\overline{Mg}]\,[Ca^{2+}]}$$

B. Mass balance:

$$[Ca]_T = [\overline{Ca}] + [Ca^{2+}]$$

$$[Mg]_T = [\overline{Mg}] + [Mg^{2+}]$$

C. Exchange capacity of the exchanger $[\overline{R}]_T$ is constant, and therefore:

$$[\overline{R}]_T = [\overline{Ca}] + [\overline{Mg}]$$

D. These equations can be rearranged into a polynomial in [Ca] by rearranging:

$$[Mg^{2+}] = [Mg]_T - [\overline{R}]_T + [Ca]_T - [Ca^{2+}]$$

$$[\overline{Mg}] = [Mg]_T - [Mg^{2+}]$$

$$[\overline{Ca}] = [Ca]_T - [Ca^{2+}]$$

E. Substituting and rearranging these equations:

$$A[Ca^{2+}]^2 + B[Ca^{2+}] + C = 0$$

where:

$$A = (K^{Ca}_{Mg} - 1)$$

$$B = [K^{Ca}_{Mg} - 1][\overline{R}]_T - [K^{Ca}_{Mg} - 2][Ca]_T + [Mg]_T$$

$$C = \left(([\overline{R}]_T - [Mg]_T) - [Ca]_T\right)[Ca]_T$$

Derivation of the Ca-Na Exchange Function

A. Selectivity coefficient:

$$K^{Ca}_{Na} = \frac{[\overline{Ca}]\,[Na^+]^2}{[\overline{Na}]^2\,[Ca^{2+}]}$$

B. Mass balance:

$$[Ca]_T = [\overline{Ca}] + [Ca^{2+}]$$

$$[Na]_T = [\overline{Na}] + [Na^+]$$

c. Exchange capacity of the exchanger $[\overline{R}]_T$ is constant, and combined with the electrical neutrality condition:

$$[\overline{R}]_T = 2[\overline{Ca}] + [\overline{Na}]$$

D. These equations can be rearranged into a polynomial in [Na]. This is arbitrary. It could also be rearranged into a polynomial in any of the other variables. Equation $[Ca]_T = [\overline{Ca}] + [Ca^{2+}]$, $[Na]_T = [\overline{Na}] + [Na^+]$, and $[\overline{R}]_T = 2[\overline{Ca}] + [\overline{Na}]$ are first rearranged into:

$$[\overline{Ca}] = \frac{[\overline{R}]_T - [\overline{Na}]}{2}$$

$$[Na^+] = [Na]_T - [\overline{Na}]$$

$$[Ca^{2+}] = [Ca]_T - 0.5[\overline{R}]_T + 0.5[\overline{Na}]$$

E. Substituting above equations into $K_{Na}^{Ca} = \dfrac{[\overline{Ca}][Na^+]^2}{[\overline{Na}]^2[Ca^{2+}]}$ and rearranging:

$$A[\overline{Na}]^3 + B[\overline{Na}] + C[\overline{Na}] + D = 0$$

where:

$$A = 0.5(K_{Na}^{Ca} + 1)$$

$$B = K_{Na}^{Ca}[Ca]_T - 0.5K_{Na}^{Ca}[\overline{R}]_T - 0.5[\overline{R}]_T - [Na]_T$$

$$C = [\overline{R}]_T[Na]_T + 0.5[Na]_T^2$$

$$D = 0.5[\overline{R}]_T[Na]_T^2$$

Ca-Mg Exchange by Farmer Fillipi

Behavior in Plates 1 and 2

Plate 1				
Water Packet#	CaX	MgX	[Ca]	[Mg]
1	300	200	250	500

2	321.8	178.2	28.2	46.8
3	340.6	159.4	31.2	43.8
4	356.6	143.4	34.0	41.0
.				
•	428.6	71.4	50	25

Plate 1				
Water Packet#	CaX	MgX	[Ca]	[Mg]
0	300	200	250	500.
1	300	200	250	500
2	302.8	197.2	25.39	49.6
3	307.9	192.1	26.1	48.9
.				
•	428.6	71.4	50	25

Behavior of Water Packet #2				
Plate #	CaX	MgX	[Ca]	[Mg]
0			50	25
1	321.8	178.2	28.2	46.8
2	302.8	197.2	25.4	49.6
3	300.3	199.7	25.04	49.96
.				
•	300	200	25	50.00

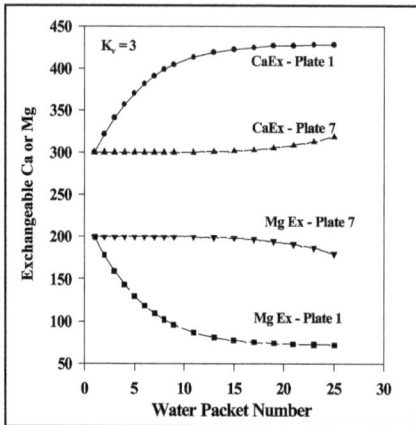

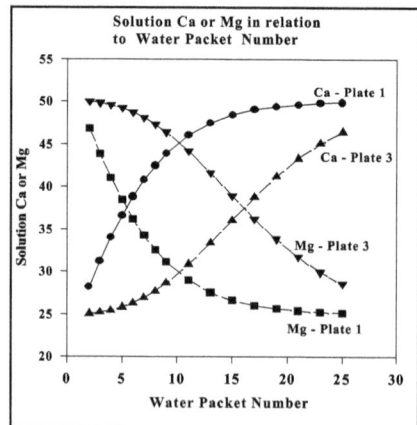

The figure above illustrates the general concepts of a cation distribution around a charged particle. Assuming that the particle is negatively charged, anions are repelled from the surface and cations are attracted to the surface. This attraction and repulsion extend outward into the solution bathing soil colloids to a greater or lesser extent depending on the charge on the colloid, the ionic strength of the bathing solution, the nature of the interaction between the cations, the anions and the surface, and the valence of the cations and anions. At come distance from the colloid, the field of the colloid is not expressed and the repulsion and attraction of cations and anions is zero. This point is called bulk solution. Since electrical neutrality must be maintained in bulk solution the charge in any volume element from negative ions is equal to the charge in this volume element from positively charged ions.

Numerically this is: $z_i^+ m_i^+ = z_i^- m_i^-$ and if we use the politically incorrect notation of equivalents then $z_i m_i = n_i$ and $n_i^+ = n_i^-$. If we use the notation that in the bulk solution $n = n_o$, then relative to n_o, n^+ increases as we approach the negative colloid and n^- decreases.

This behavior is analogous too and well illustrated by for the distribution of gases in a gravitational field. To help us develop a feel for the colloidal system, we will examine the behavior of gases in the earths gravitational field. In this case gases decrease in relation to the distance from the earth's surface. If we designate n_o as the value of the gas concentration at the earth's surface, we can develop a model of the gas distribution.

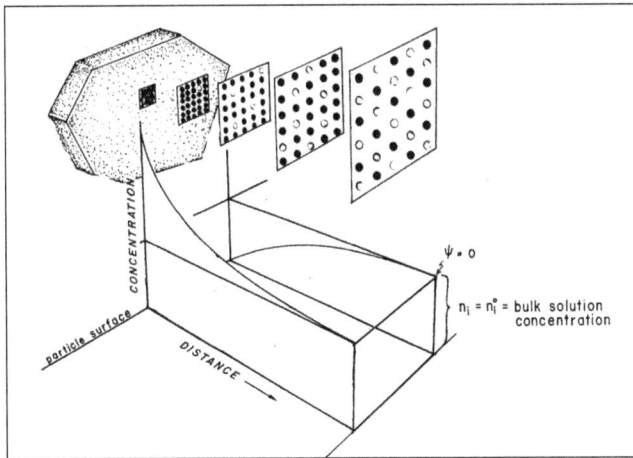

Barometric Formula and Analogy

$$n = n_o \exp\left(-\frac{Mgh}{kT}\right)$$

where,

    n = the number of gas atoms per cubic centimeter at height h

    n_o = the number of gas atoms per cubic centimeter at h = 0, (earth's surface)

    m = the mass of the atoms

g = the gravitational constant

k = the Boltzmann constant

T = absolute temperature (K_o)

mgh = the gravitational potential of the gas atoms

kT = the thermal or kinetic energy of the gas atoms (escaping energy)

Thus gases are distributed in around the earth in relation to their thermal energy (kT) or escaping tendency and the gravitational attraction (mgh). By a similar analogy, ions are distributed in an electrical field around a charged colloid in relation to two opposing forces.

Boltzmann Expression for Ions in an Electrical Field

$$n_+ = n_o \, exp-\left(\frac{z_+ \; c \, \psi}{kT} \right)$$

$$n_- = n_o \, exp\left(\frac{|z_-| \; c \, \psi}{kT} \right)$$

n_+ = local cation concentration

: ions cm^{-3}

n_-= local anion concentration

: ions cm^{-3}

n_o = concentration of cations and anions

: ions cm^{-3}

z_+ = cation valence

: electrons ion^{-1}

½z½= absolute value of the anion valence

: electrons ion^{-1}

c = the charge on the electron = 1 unit of charge =

4.8 * 10^{-10} statvolts electron^{-1}

n_i = local electrical potential, the sign of i is negative for negatively charged particles.

: statvolts

Note 1 statvolt = 300 volts

$z_i c n_i$ = electrical potential of the ion

: ergs ion^{-1}

Note: 1 statvolt* statcoulomb = 1 erg

Define a convenience parameter (y):

$$y = \left(\frac{z_i \; c \; \psi}{kT} \right)$$

then from $n_+ = n_\circ \exp - \left(\frac{z_+ \; c \; \psi}{kT} \right)$ and $n_- = n_\circ \exp \left(\frac{|z_-| \; c \; \psi}{kT} \right)$:

$$n^+ = n_\circ \exp(-y) \quad \text{or} \quad n_+ = n_\circ e^{-y}$$

$$n_- = n_\circ \exp(y) \quad \text{or} \quad n_- = n_\circ e^{y}$$

Introduce the Poisson equation from classical electrostatic theory:

$$\frac{d^2 \psi}{dx^2} = -\frac{4 \pi p}{D}$$

x = the normal distance from a flat surface

D = the dielectric constant

: statcoulombs cm^{-1} statvolts^{-1}

d = local net charge density

p = is defined as the difference in charge in any volume element found by summing the positive and negative charges in that volume element.

$$= \Sigma \left(z_i c \; n_i \right)$$

: statcoulombs cm^{-3}

substituting for *p* in $\frac{d^2 \psi}{dx^2} = -\frac{4 \pi p}{D}$ yields:

$$\frac{d^2 \psi}{dx^2} = -\frac{4 \pi}{D} \Sigma z_i c \; n_\circ \exp\left(-\frac{z_i \; c \psi}{kT} \right)$$

If we assume that the solution contains a single symmetrical electrolyte we may rewrite
$\dfrac{d^2\psi}{dx^2} = -\dfrac{4\pi}{D}\Sigma z_i c\; n_o\; \exp\left(-\dfrac{z_i\; c\psi}{kT}\right)$ to give:

$$\frac{d^2\psi}{dx^2} = -\frac{4\pi\; |z_i|\, c\; n_o}{D}\left[\exp\left(-\frac{|z_i|\; n_o\psi}{kT}\right) - \exp\left(-\frac{|z_i|\, c\psi}{kT}\right)\right]$$

Introducing y from $y = \left(\dfrac{z_i\;\; c\;\psi}{kT}\right)$ after finding $\dfrac{d^2 y}{dx^2}$ and recalling that:

$\exp(y) - \exp(-y)\; \sinh\; y = \dfrac{\exp(y) - \exp(-y)}{2}$, we can rewrite:

$$\frac{d^2\psi}{dx^2} = -\frac{4\pi\; |z_i|\, c\; n_o}{D}\left[\exp\left(-\frac{|z_i|\; n_o\psi}{kT}\right) - \exp\left(-\frac{|z_i|\, c\psi}{kT}\right)\right]\; \text{as:}$$

$$\frac{d^2 y}{dx^2} = \frac{8\pi\; z_1^2\; c^2\; n_o\; \sinh(y)}{DkT}$$

Since there are a number of constants in equation $\dfrac{d^2 y}{dx^2} = \dfrac{8\pi\; z_1^2\; c^2\; n_o\; \sinh(y)}{DkT}$, it is convenient to introduce the convenience parameter kappa squared:

$$k^2 = \frac{8\pi\; z_1^2\; c^2\; n_o}{DkT}$$

having units of cm⁻¹

$$\frac{d^2 y}{dx^2} = k^2\; \sinh(y)$$

Applying an integrating factor to both sides and using the boundary conditions that at $x = \infty$, $y = 0$ and $dy/dx = 0$ it can be shown:

$$\frac{dy}{dx} = 2k\; \sinh\left(\frac{y}{2}\right)$$

Adding that $y = y_o$ at $x = 0$, and after some manipulations $\dfrac{dy}{dx} = 2k\; \sinh\left(\dfrac{y}{2}\right)$ may be integrated to the working equation:

$$y = 2\ln\left[\frac{a + \exp(kx)}{-a + \exp(kx)}\right]$$

Where, $a = \tanh\left(\dfrac{y_\circ}{2}\right)$ and $y_\circ = \left(\dfrac{|z_i|\,c\,\psi_\circ}{kT}\right)$

We proceed by introducing the space charge (s) which is equal and opposite to the surface charge, and has units of statvolts cm^{-2}:

$$\sigma = \int_0^\infty \rho\ dx$$

Substituting for (ρ) from $\dfrac{d^2\psi}{dx^2} = -\dfrac{4\pi p}{D}$ into equation $\sigma = \int_0^\infty \rho\ dx$ yields:

$$\sigma = -\int_0^\infty \dfrac{D}{4\pi}\dfrac{d^2\psi}{dx^2}dx$$

which after some manipulations integrates to:

$$\sigma = -\dfrac{4c|z_i|n_\circ}{k}\sinh\left(\dfrac{y_\circ}{2}\right)$$

If we again introduce a convenience parameter J defined as:

$$J = \dfrac{4|z_i|c}{\left(\dfrac{8\pi\ z_i^2\ c^2}{DkT}\right)^{0.5}}$$

introducing the constant makes it easier to examine the relation between (s) and surface potential. Remember that surface potential is contained in the yo term.

$$\sigma = -J\sqrt{n_\circ}\ \sinh\left(\dfrac{y_\circ}{2}\right)$$

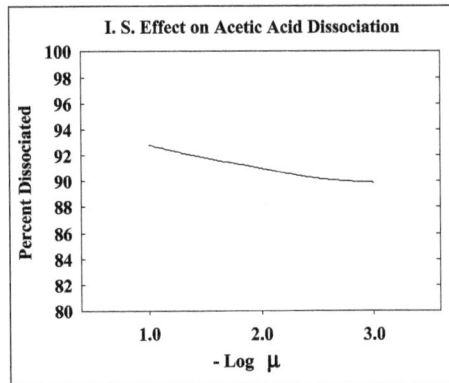

I. S. Effect on Acetic Acid Dissociation

It has been shown that for variable charge surfaces:

$$\psi_\circ = \dfrac{kT}{z\ c}\ln\left(\dfrac{a_{PDI}}{a_{PDIzpc}}\right)$$

Where,

ψ_0 = surface potential of the colloid

a_{PDI} = solution activity of the potential determiningion

a_{PDIzpc} = the solution activity of the potential determining ion at the point of zero charge (zpc)

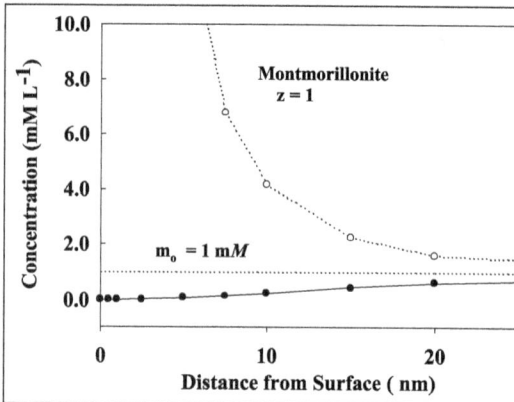

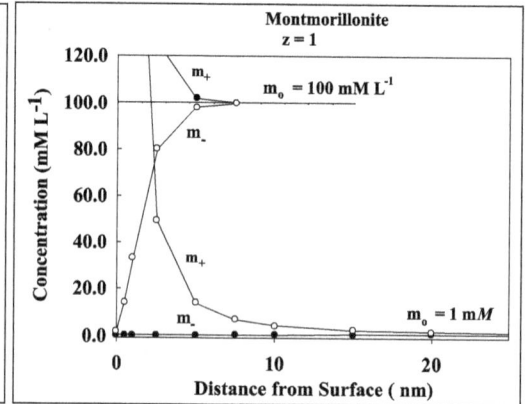

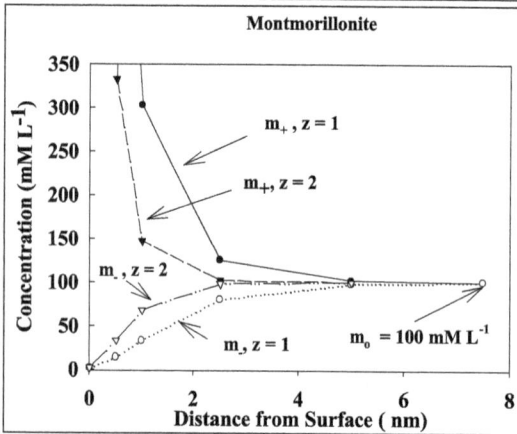

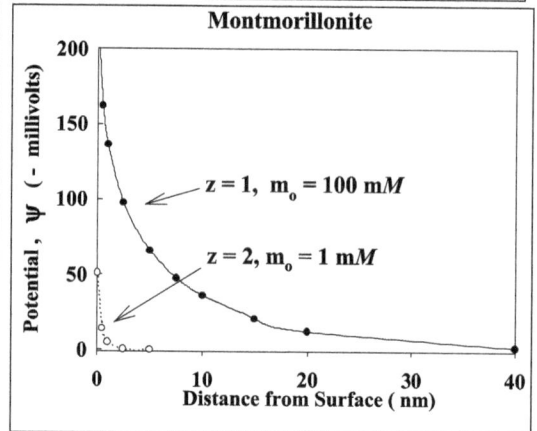

References

- A-comp, mauisoil: ctahr.hawaii.edu, Retrieved 13 May, 2019

- Mosley, l.m.; zammit, b.; jolley, a.m.; barnett, l. (2014). "acidification of lake water due to drought". Journal of hydrology. 511: 484–493. Bibcode:2014jhyd..511..484m. Doi:10.1016/j.jhydrol.2014.02.001

- Phosphoric-acid-h3po4, definition: maximumyield.com, Retrieved 25 February, 2019

- Mosley, lm; zammit, b; jolley, a; barnett, l; fitzpatrick, r (2014). "monitoring and assessment of surface water acidification following rewetting of oxidised acid sulfate soils". Environmental monitoring and assessment. 186: 1–18. Doi:10.1007/s10661-013-3350-9. Pmid 23900634

- Calcium-in-the-soil, indepth-articles, crops: growersmineral.com, Retrieved 16 January, 2019

4

Manures and Fertilizers

The organic matter which is generally derived from animal feces is termed as manure. Any material of synthetic or natural origin which is applied to soil or plant tissues to aid the growth of plants is termed as a fertilizer. All the diverse principles of manures and fertilizers as well as their different types have been carefully analyzed in this chapter.

MANURE

Manure is an organic material that is used to fertilize land, usually consisting of the feces and urine of domestic livestock, with or without accompanying litter such as straw, hay, or bedding. Farm animals void most of the nitrogen, phosphorus, and potassium that is present in the food they eat, and this constitutes an enormous fertility resource. In some countries, human excrement is also used. Livestock manure is less rich in nitrogen, phosphorus, and potash than synthetic fertilizers and hence must be applied in much greater quantities than the latter. A ton of manure from cattle, hogs, or horses usually contains only 10 pounds of nitrogen, 5 pounds of phosphorus pentoxide, and 10 pounds of potash. But manure is rich in organic matter, or humus, and thus improves the soil's capacity to absorb and store water, thus preventing erosion. Much of the potassium and nitrogen in manure can be lost through leaching if the material is exposed to rainfall before being applied to the field. These nutrient losses may be prevented by such methods as stacking manure under cover or in pits to prevent leaching, spreading it on fields as soon as it is feasible, and spreading preservative materials in the stable. A green manure is a cover crop of some kind, such as rye, that is plowed under while still green to add fertility and conditioning to the soil.

The use of manure as fertilizer dates to the beginnings of agriculture. On modern farms manure is usually applied with a manure spreader, a four-wheeled self-propelled or two-wheeled tractor-drawn wagon. As the spreader moves, a drag-chain conveyor located at the bottom of the box sweeps the manure to the rear, where it is successively shredded by a pair of beaters before being spread by rotating spiral fins. Home gardeners like to use well-rotted manure, since it is less odorous, more easily spread, and less likely to "burn" plants.

GREEN MANURE

In agriculture, green manure is created by leaving uprooted or sown crop parts to wither on a field so that they serve as a mulch and soil amendment. The plants used for green manure are often cover crops grown primarily for this purpose. Typically, they are ploughed under and incorporated into the soil while green or shortly after flowering. Green manure is commonly associated with organic farming and can play an important role in sustainable annual cropping systems.

Functions

Green manures usually perform multiple functions that include soil improvement and soil protection.

Leguminous green manures such as clover and vetch contain nitrogen-fixing symbiotic bacteria in root nodules that fix atmospheric nitrogen in a form that plants can use. This performs the vital function of fertilization. If desired, animal manures may also be added.

Depending on the species of cover crop grown, the amount of nitrogen released into the soil lies between 40 and 200 pounds per acre. With green manure use, the amount of nitrogen that is available to the succeeding crop is usually in the range of 40-60% of the total amount of nitrogen that is contained within the green manure crop.

Average biomass yields and nitrogen yields of several legumes by crop	Biomass tons acre^{-1}	N lbs acre^{-1}
Sweet clover	1.75	120
Berseem clover	1.10	70
Crimson clover	1.40	100
Hairy vetch	1.75	110

- Green manure acts mainly as soil-acidifying matter to decrease the alkalinity/pH of alkali soils by generating humic acid and acetic acid.

- Incorporation of cover crops into the soil allows the nutrients held within the green manure to be released and made available to the succeeding crops. This results immediately from an increase in abundance of soil microorganisms from the degradation of plant material that aid in the decomposition of this fresh material. This additional decomposition also allows for the re-incorporation of nutrients that are found in the soil in a particular form such as nitrogen (N), potassium (K), phosphorus (P), calcium (Ca), magnesium (Mg), and sulfur (S).

- Microbial activity from incorporation of cover crops into the soil leads to the formation of mycelium and viscous materials which benefit the health of the soil by increasing its soil structure (i.e. by aggregation).

The increased percentage of organic matter (biomass) improves water infiltration and retention, aeration, and other soil characteristics. The soil is more easily turned or tilled than non-aggregated soil. Further aeration of the soil results from the ability of the root systems of many green manure crops to efficiently penetrate compact soils. The amount of humus found in the soil also increases with higher rates of decomposition, which is beneficial for the growth of the crop succeeding the green manure crop. Non-leguminous crops are primarily used to increase biomass.

- The root systems of some varieties of green manure grow deep in the soil and bring up nutrient resources unavailable to shallower-rooted crops.

- Common cover crop functions of weed suppression. Non-leguminous crops are primarily used (e.g. buckwheat) The deep rooting properties of many green manure crops make them efficient at suppressing weeds.

- Some green manure crops, when allowed to flower, provide forage for pollinating insects. Green manure crops also often provide habitat for predatory beneficial insects, which allow for a reduction in the application of insecticides where cover crops are planted.

- Some green manure crops (e.g. winter wheat and winter rye) can also be used for grazing.

- Erosion control is often also taken into account when selecting which green manure cover crop to plant.

- Some green crops reduce plant insect pests and diseases. Verticillium wilt is especially reduced in potato plants.

Incorporation of green manures into a farming system can drastically reduce the need for additional products such as supplemental fertilizers and pesticides.

Limitations to consider in the use of green manure are time, energy, and resources (monetary and natural) required to successfully grow and utilize these cover crops. Consequently, it is important to choose green manure crops based on the growing region and annual precipitation amounts to ensure efficient growth and use of the cover crop(s).

Nutrient Creation

Green manure is broken down into plant nutrient components by heterotrophic bacteria that consumes organic matter. Warmth and moisture contribute to this process, similar to creating compost fertilizer. The plant matter releases large amounts of carbon dioxide and weak acids that react with insoluble soil minerals to release beneficial nutrients. Soils that are high in calcium minerals, for example, can be given green manure to generate a higher phosphate content in the soil, which in turn acts as a fertilizer.

The ratio of carbon to nitrogen in a plant is a crucial factor to consider, since it will impact the nutrient content of the soil and may starve a crop of nitrogen, if the incorrect plants are used to make green manure. The ratio of carbon to nitrogen will differ from species to species, and depending upon the age of the plant. The ratio is referred to as C:N. The value of N is always one, whereas the value of carbon or carbohydrates is expressed in a value of about 10 up to 90; the ratio must be less than 30:1 to prevent the manure bacteria from depleting existing nitrogen in the soil. Rhizobium are soil organisms that interact with green manure to retain atmospheric nitrogen in the soil. Legumes, such as beans, alfalfa, clover and lupines, have root systems rich in rhizobium, often making them the preferred source of green manure material.

Crops

Late-summer and fall green manure crops are oats and rye.

Other green manure crops:

- Alfalfa, which sends roots deep to bring nutrients to the surface,
- Buckwheatin temperate regions,
- Cowpea,
- Clover (e.g. annual sweet clover),
- Fava beans,
- Fenugreek,
- Ferns of the genus *Azolla* have been used as a green manure in southeast Asia,
- Lupin,
- Groundnut,
- Millet,
- Mustard,
- *Phacelia tanacetifolia,*
- Radish such as tillage radish or daikon radish,
- Sesbania,
- Sorghum,
- Soybean,
- Sudangrass,

- Sunn hemp, a legume widely grown throughout the tropics and subtropics,

- Tyfon, a *Brassica* known for a strong tap root that breaks up heavy soils,

- Velvet bean (*Mucuna pruriens*), common in the southern US during the early part of the 20th century, before being replaced by soybeans, popular today in most tropical countries, especially in Central America, where it is the main green manure used in slash/mulch farming practices,

- Vetch (*Vicia sativa, Vicia villosa*):

  ○ This is one of many legumes that may be used as a green manure crop.

  ○ This is one of many non-legumes that may be used as a green manure crop.

  ○ This is one of many legumes that may be used as a green manure crop.

FARMYARD MANURE

Farmyard Manure is prepared basically using cow dung, cow urine, waste straw and other dairy wastes. It is highly useful and some of its properties are given below:

- FYM is rich in nutrients.

- A small portion of N is directly available to the plants while a larger portion is made available as and when the FYM decomposes.

- When cow dung and urine are mixed, a balanced nutrition is made available to the plants.

- Availability of Potassium and Phosphorus from FYM is similar to that from inorganic sources.

- Application of FYM improves soil fertility.

Nutritional status of FYM (%):

Nitrogen	0.5000
Phosphorus	0.2500
Potassium	0.4000
Calcium	0.0800
Sulfur	0.0200
Zinc	0.0040
Copper	0.0003
Manganese	0.0070
Iron	0.4500

Cow dung which we get in abundance (10 cows) is collected after cleaning cowshed in a pit close by and is allowed to decompose over a period of time. Every month this manure (compost) is applied to the plants or the field to enrich the soil.

Since we use almost the entire cow dung for Gobar gas plant, the slurry that comes out after gas generation is highly composted and extremely rich.

CHICKEN MANURE

Chicken manure is the feces of chickens used as an organic fertilizer, especially for soil low in nitrogen. Of all animal manures, it has the highest amount of nitrogen, phosphorus, and potassium. Chicken manure is sometimes pelletized for use as a fertilizer, and this product may have additional phosphorus, potassium or nitrogen added. Optimal storage conditions for chicken manure includes it being kept in a covered area and retaining its liquid, because a significant amount of nitrogen exists in the urine.

Fresh chicken manure contains 0.8% potassium, 0.4% to 0.5% phosphorus and 0.9% to 1.5% nitrogen. One chicken produces approximately 8–11 pounds of manure monthly. Chicken manure can be used to create homemade plant fertilizer.

Pollution

Mass applications of chicken manure may create an unpleasant odor. In April 2014 in Escondido, California, a golf course that had "dumped" chicken manure on its grounds was cited by the county government after complaints from local residents about the odor.

In December 2011, the environmental group Environment Maryland asserted that water runoff from agricultural land fertilized with chicken manure was increasing the pollution levels of Chesapeake Bay. The group asserted that excessive phosphorus from the runoff was contributing to the increase of dead zones in the bay, in efforts to address the matter before leaving office, Maryland Governor Martin O'Malley put a new regulation into use that "would have limited the amount of poultry manure that Eastern Shore farmers can use on their fields". However, the following Governor Larry Hogan quickly absolved the new regulation after being sworn into office. The runoff problem has been attributed to the use of "an outdated scientific tool for calculating the correct amount of manure". A proposed solution from scientists at the University of Maryland is to have farmers use a new (corrected) formula to calculate proper quantities of chicken manure for agricultural uses.

Human Deterrent

Chicken manure has been used as a human deterrent. In July 2013, in Abbotsford, British Columbia, city workers applied chicken manure at a tent encampment to deter homeless people from the area. The affected homeless planned on initiating small claims lawsuits for loss of property and property damage. One of the affected homeless people described the tactics of city workers as "a chicken shit way to do things". The mayor of Abbotsford and the Fraser Valley city manager later apologized regarding the incident. Similar instances of using chicken manure in this manner have occurred in British Columbia in Surrey and in Port Coquitlam, the latter of which occurred "shortly after the Abbotsford incident".

COMPOST

Compost is organic matter that has been decomposed in a process called composting. This process recycles various organic materials otherwise regarded as waste products and produces a soil conditioner (the *compost*).

Compost is rich in nutrients. It is used, for example, in gardens, landscaping, horticulture, urban agriculture and organic farming. The compost itself is beneficial for the land in many ways, including as a soil conditioner, a fertilizer, addition of vital humus or humic acids, and as a natural pesticide for soil. In ecosystems, compost is useful for erosion control, land and stream reclamation, wetland construction, and as landfill cover.

At the simplest level, the process of composting requires making a heap of wet organic matter (also called green waste), such as leaves, grass, and food scraps, and waiting for the materials to break down into humus after a period of months. However, composting also can take place as a multi-step, closely monitored process with measured inputs of water, air, and carbon- and nitrogen-rich materials. The decomposition process is aided by shredding the plant matter, adding water and ensuring proper aeration by regularly turning the mixture when open piles or "windrows" are used. Earthworms and fungi further break up the material. Bacteria requiring oxygen to function (aerobic bacteria) and fungi manage the chemical process by converting the inputs into heat, carbon dioxide, and ammonium.

Fundamentals

Home compost barrel.   Materials in a compost pile.   Food scraps compost heap.

Composting is an aerobic method (meaning that it requires the presence of air) of decomposing organic solid wastes. It can therefore be used to recycle organic material. The process involves decomposition of organic material into a humus-like material, known as compost, which is a good fertilizer for plants. Composting requires the following three components: human management, aerobic conditions, development of internal biological heat.

Composting organisms require four equally important ingredients to work effectively:

- Carbon — For energy; the microbial oxidation of carbon produces the heat, if included at suggested levels. High carbon materials tend to be brown and dry.

- Nitrogen — To grow and reproduce more organisms to oxidize the carbon. High nitrogen materials tend to be green (or colorful, such as fruits and vegetables) and wet.

- Oxygen — For oxidizing the carbon, the decomposition process.

- Water — In the right amounts to maintain activity without causing anaerobic conditions.

Certain ratios of these materials will provide microorganisms to work at a rate that will heat up the pile. Active management of the pile (e.g. turning) is needed to maintain sufficient supply of oxygen and the right moisture level. The air/water balance is critical to maintaining high temperatures (135°-160° Fahrenheit / 50° - 70° Celsius) until the materials are broken down.

The most efficient composting occurs with an optimal carbon:nitrogen ratio of about 25:1. Hot container composting focuses on retaining the heat to increase decomposition rate and produce compost more quickly. Rapid composting is favored by having a

C/N ratio of ~30 or less. Above 30 the substrate is nitrogen starved, below 15 it is likely to outgas a portion of nitrogen as ammonia.

Nearly all plant and animal materials have both carbon and nitrogen, but amounts vary widely, with characteristics noted above (dry/wet, brown/green). Fresh grass clippings have an average ratio of about 15:1 and dry autumn leaves about 50:1 depending on species. Mixing equal parts by volume approximates the ideal C:N range. Few individual situations will provide the ideal mix of materials at any point. Observation of amounts, and consideration of different materials as a pile is built over time, can quickly achieve a workable technique for the individual situation.

Microorganisms

With the proper mixture of water, oxygen, carbon, and nitrogen, micro-organisms are able to break down organic matter to produce compost. The composting process is dependent on micro-organisms to break down organic matter into compost. There are many types of microorganisms found in active compost of which the most common are:

- Bacteria - The most numerous of all the microorganisms found in compost. Depending on the phase of composting, mesophilic or thermophilic bacteria may predominate.

- Actinobacteria - Necessary for breaking down paper products such as newspaper, bark, etc.

- Fungi - Molds and yeast help break down materials that bacteria cannot, especially lignin in woody material.

- Protozoa - Help consume bacteria, fungi and micro organic particulates.

- Rotifers - Rotifers help control populations of bacteria and small protozoans.

In addition, earthworms not only ingest partly composted material, but also continually re-create aeration and drainage tunnels as they move through the compost.

Phases of Composting

Three years old household compost.

Under ideal conditions, composting proceeds through three major phases:

- An initial, mesophilic phase, in which the decomposition is carried out under moderate temperatures by mesophilic microorganisms.

- As the temperature rises, a second, thermophilic phase starts, in which the decomposition is carried out by various thermophilic bacteria under high temperatures.

- As the supply of high-energy compounds dwindles, the temperature starts to decrease, and the mesophiles once again predominate in the maturation phase.

Slow and Rapid Composting

There are many proponents of rapid composting that attempt to correct some of the perceived problems associated with traditional, slow composting. Many advocate that compost can be made in 2 to 3 weeks. Many such short processes involve a few changes to traditional methods, including smaller, more homogenized pieces in the compost, controlling carbon-to-nitrogen ratio (C:N) at 30 to 1 or less, and monitoring the moisture level more carefully. However, none of these parameters differ significantly from the early writings of compost researchers, suggesting that in fact modern composting has not made significant advances over the traditional methods that take a few months to work. For this reason and others, many scientists who deal with carbon transformations are sceptical that there is a "super-charged" way to get nature to make compost rapidly.

Both sides may be right to some extent. The bacterial activity in rapid high heat methods breaks down the material to the extent that pathogens and seeds are destroyed, and the original feedstock is unrecognizable. At this stage, the compost can be used to prepare fields or other planting areas. However, most professionals recommend that the compost be given time to cure before using in a nursery for starting seeds or growing young plants. The curing time allows fungi to continue the decomposition process and eliminating phytotoxic substances.

An alternative approach is anaerobic fermentation, known as bokashi. It retains carbon bonds, is faster than decomposition, and for application to soil requires only rapid but thorough aeration rather than curing. It depends on sufficient carbohydrates in the treated material.

Pathogen Removal

Composting can destroy pathogens or unwanted seeds. Unwanted living plants (or weeds) can be discouraged by covering with mulch/compost. The "microbial pesticides" in compost may include thermophiles and mesophiles.

Thermophilic (high-temperature) composting is well known to destroy many seeds and nearly all types of pathogens (exceptions may include prions). The sanitizing qualities of (thermophilic) composting are desirable where there is a high likelihood of pathogens, such as with manure.

Materials that can be Composted

Composting is a process used for resource recovery. It can recycle an unwanted by-product from another process (a waste) into a useful new product.

Organic Solid Waste (Green Waste)

A large compost pile that is steaming with the heat generated by thermophilic microorganisms.

Composting is a process for converting decomposable organic materials into useful stable products. Therefore, valuable landfill space can be used for other wastes by composting these materials rather than dumping them on landfills. It may however be difficult to control inert and plastics contamination from municipal solid waste.

Co-composting is a technique that processes organic solid waste together with other input materials such as dewatered fecal sludge or sewage sludge.

Industrial composting systems are being installed to treat organic solid waste and recycle it rather than landfilling it. It is one example of an advanced waste processing system. Mechanical sorting of mixed waste streams combined with anaerobic digestion or in-vessel composting is called mechanical biological treatment. It is increasingly being used in developed countries due to regulations controlling the amount of organic matter allowed in landfills. Treating biodegradable waste before it enters a landfill reduces global warming from fugitive methane; untreated waste breaks down anaerobically in a landfill, producing landfill gas that contains methane, a potent greenhouse gas.

Animal Manure and Bedding

On many farms, the basic composting ingredients are animal manure generated on the farm and bedding. Straw and sawdust are common bedding materials. Non-traditional bedding materials are also used, including newspaper and chopped cardboard. The amount of manure composted on a livestock farm is often determined by cleaning schedules, land availability, and weather conditions. Each type of manure has its own physical, chemical, and biological characteristics. Cattle and horse manures, when mixed with bedding, possess good qualities for composting. Swine manure, which is very wet and usually not mixed with bedding material, must be mixed with straw or similar raw materials. Poultry manure also must be blended with carbonaceous materials - those low in nitrogen preferred, such as sawdust or straw.

Human Excreta and Sewage Sludge

Human excreta can also be added as an input to the composting process since human excreta is a nitrogen-rich organic material. It can be either composted directly, like in composting toilets, or indirectly (as sewage sludge), after it has undergone treatment in a sewage treatment plant.

Urine can be put on compost piles or directly used as fertilizer. Adding urine to compost can increase temperatures and therefore increase its ability to destroy pathogens and unwanted seeds. Unlike feces, urine does not attract disease-spreading flies (such as houseflies or blowflies), and it does not contain the most hardy of pathogens, such as parasitic worm eggs. Urine usually does not smell for long, particularly when it is fresh, diluted, or put on sorbents.

Uses

Compost can be used as an additive to soil, or other matrices such as coir and peat, as a tilth improver, supplying humus and nutrients. It provides a rich *growing medium*, or a porous, absorbent material that holds moisture and soluble minerals, providing the support and nutrients in which plants can flourish, although it is rarely used alone, being primarily mixed with soil, sand, grit, bark chips, vermiculite, perlite, or clay granules to produce loam. Compost can be tilled directly into the soil or growing medium to boost the level of organic matter and the overall fertility of the soil. Compost that is ready to be used as an additive is dark brown or even black with an earthy smell.

Generally, direct seeding into a compost is not recommended due to the speed with which it may dry and the possible presence of phytotoxins in immature compost that may inhibit germination, and the possible tie up of nitrogen by incompletely decomposed lignin. It is very common to see blends of 20–30% compost used for transplanting seedlings at cotyledon stage or later.

Compost can be used to increase plant immunity to diseases and pests.

Composting Technologies

Various approaches have been developed to handle different ingredients, locations, throughput and applications for the composted product.

Industrial-scale

Industrial-scale composting can be carried out in the form of in-vessel composting, aerated static pile composting, vermicomposting, or windrow composting.

Vermicomposting

Worms in a bin being harvested.

Vermicompost is the product or process of organic material degradation using various species of worms, usually red wigglers, white worms, and earthworms, to create a heterogeneous mixture of decomposing vegetable or food waste (excluding meat, dairy, fats, or oils), bedding materials, and vermicast. Vermicast, also known as worm castings, worm humus or worm manure, is the end-product of the breakdown of organic matter by species of earthworm. Vermicomposting can also be applied for treatment of sewage sludge.

Composting Toilets

Composting toilet with a seal in the lid.

A composting toilet collects human excreta. These are added to a compost heap that can be located in a chamber below the toilet seat. Sawdust and straw or other carbon rich materials are usually added as well. Some composting toilets do not require water or electricity; others may. If they do not use water for flushing they fall into the category of dry toilets. Some composting toilet designs use urine diversion, others do not. When properly managed, they do not smell. The composting process in these toilets destroys pathogens to some extent. The amount of pathogen destruction depends on the temperature (mesophilic or thermophilic conditions) and composting time.

Composting toilets with a large composting container (of the type Clivus Multrum and derivations of it) are popular in United States, Canada, Australia, New Zealand and Sweden. They are available as commercial products, as designs for self builders or as "design derivatives" which are marketed under various names.

Black Soldier Fly Larvae

Black soldier fly (*Hermetia illucens*) larvae are able to rapidly consume large amounts of organic material when kept at around 30 °C. Black soldier fly larvae can reduce the dry matter of the organic waste by 73% and convert 16-22% of the dry matter in the waste to biomass. The resulting compost still contains nutrients and can be used for biogas production, or further traditional composting or vermicomposting. The larvae are rich in fat and protein, and can be used as, for example, animal feed or biodiesel production. Enthusiasts have experimented with a large number of different waste products.

Bokashi

Bokashi is not composting as defined earlier, rather an alternative technology. It ferments (rather than decomposes) the input organic matter and feeds the result to the soil food web (rather than producing a soil conditioner). The process involves adding *Lactobacilli* to the input in an airtight container kept at normal room temperature. These bacteria ferment carbohydrates to lactic acid, which preserves the input. After this is complete the preserve is mixed into soil, converting the lactic acid to pyruvate, which enables soil life to consume the result.

Bokashi is typically applied to food waste from households, workplaces and catering establishments, because such waste normally holds a good proportion of carbohydrates; it is also applied to other organic waste by supplementing carbohydrates. Household containers ("bokashi bins") typically give a batch size of 5-10 kilograms, accumulated over a few weeks. In horticultural settings batches can be orders of magnitude greater.

Inside a recently started bokashi bin. Food scraps are raised on a perforated plate (to drain runoff) and are partly covered by a layer of bran inoculated with *Lactobacilli*.

Bokashi offers several advantages:

- Fermentation retains all the original carbon and energy. (In comparison, composting loses at least 50% of these and 75% or more in amateur use; composting also loses nitrogen, a macronutrient of plants, by emitting ammonia and the potent greenhouse gas nitrous oxide.)

- Virtually the full range of food waste is accepted, without the exclusions of composting. The exception is large bones.

- Being airtight, the container inherently traps smells, and when opened the smell of fermentation is far less offensive than decomposition. Hence bokashi bins usually operate indoors, in or near kitchens.

- Similarly the container neither attracts insect pests nor allows them ingress.

- The process is inherently hygienic because lactic acid is a natural bactericide and anti-pathogen; even its own fermentation is self-limiting.

- Both preservation and consumption complete within a few weeks rather than months.

- The preserve can be stored until needed, for example if ground is frozen or waterlogged.

- The increased activity of the soil food web improves the soil texture, especially by worm action - in effect this is in-soil vermicomposting.

The importance of the first advantage should not be underestimated: the mass of any ecosystem depends on the energy it captures. Plants depend upon the soil ecosystem making nutrients available within soil water. Therefore, the richer the ecosystem, the richer the plants. (Plants can also take up nutrients from added chemicals, but these are at odds with the purpose of composting).

Other Systems at Household Level

Hügelkultur (Raised Garden Beds or Mounds)

An almost completed Hügelkultur bed; the bed does not have soil on it yet.

The practice of making raised garden beds or mounds filled with rotting wood is also called hügelkultur in German. It is in effect creating a nurse log that is covered with soil.

Benefits of hügelkultur garden beds include water retention and warming of soil.[35] Buried wood acts like a sponge as it decomposes, able to capture water and store it for later use by crops planted on top of the hügelkultur bed.

Compost Tea

Compost teas are defined as water extracts leached from composted materials. Compost teas are generally produced from adding one volume of compost to 4-10 volumes of water, but there has also been debate about the benefits of aerating the mixture. Field studies have shown the benefits of adding compost teas to crops due to the adding of organic matter, increased nutrient availability and increased microbial activity. They have also been shown to have an effect on plant pathogens.

Worm Hotels

Worm Hotel.

Worm Hotels accommodate useful worm in ideal conditions.

Related Technologies

Organic ingredients intended for composting can also be used to generate biogas through anaerobic digestion. This process stabilizes organic material. The residual material, sometimes in combination with sewage sludge can be treated by a composting process before selling or giving away the compost.

Regulations

There are process and product guidelines in Europe that date to the early 1980s and only more recently in the UK and the US. In both these countries, private trade associations within the industry have established loose standards, some say as a stop-gap measure to discourage independent government agencies from establishing tougher consumer-friendly standards.

The USA is the only Western country that does not distinguish sludge-source compost from green-composts, and by default in the USA 50% of states expect composts to comply in some manner with the federal EPA 503 rule promulgated in 1984 for sludge products. Compost is regulated in Canada and Australia as well.

Many countries such as Wales and some individual cities such as Seattle and San Francisco require food and yard waste to be sorted for composting (San Francisco Mandatory Recycling and Composting Ordinance).

Edmonton Composting Facility.

Large-scale composting systems are used by many urban areas around the world.

- The world's largest municipal co-composter for municipal solid waste (MSW) is the Edmonton Composting Facility in Edmonton, Alberta, Canada, which turns 220,000 tonnes of municipal solid waste and 22,500 dry tonnes of sewage sludge per year into 80,000 tonnes of compost. The facility is 38,690 m² (416,500 sq.ft.) in area, equivalent to 4½ Canadian football fields, and the operating structure is the largest stainless steel building in North America.

- In 2006, Qatar awarded Keppel Seghers Singapore, a subsidiary of Keppel Corporation, a contract to begin construction on a 275,000 tonne/year anaerobic digestion and composting plant licensed by Kompogas Switzerland. This plant, with 15 independent anaerobic digesters, will be the world's largest composting facility once fully operational in early 2011 and forms part of Qatar's Domestic Solid Waste Management Centre, the largest integrated waste management complex in the Middle East.

- Another large municipal solid waste composter is the Lahore Composting Facility in Lahore, Pakistan, which has a capacity to convert 1,000 tonnes of municipal solid waste per day into compost. It also has a capacity to convert substantial portion of the intake into refuse-derived fuel (RDF) materials for further combustion use in several energy consuming industries across Pakistan, for example in cement manufacturing companies where it is used to heat cement kilns. This project has also been approved by the Executive Board of the United Nations Framework Convention on Climate Change for reducing methane emissions, and has been registered with a capacity of reducing 108,686 tonnes carbon dioxide equivalent per annum.

- Kew Gardens in London has one of the biggest non-commercial compost heaps in Europe.

- Compost is used as a soil amendment in organic farming.

Vermicompost

Vermicompost (vermi-compost, vermiculture) is the product of the composting process using various species of worms, usually red wigglers, white worms, and other earthworms, to create a mixture of decomposing vegetable or food waste, bedding materials, and vermicast.

Vermicast (also called worm castings, worm humus, worm manure, or worm feces) is the end-product of the breakdown of organic matter by earthworms. These castings have been shown to contain reduced levels of contaminants and a higher saturation of nutrients than the organic materials before vermicomposting.

Vermicompost contains water-soluble nutrients and is an excellent, nutrient-rich organic fertilizer and soil conditioner. It is used in farming and small scale sustainable, organic farming.

Vermicomposting can also be applied for treatment of sewage. A variation of the process is vermifiltration (or vermidigestion) which is used to remove organic matter, pathogens and oxygen demand from wastewater or directly from blackwater of flush toilets.

Vermicomposting has gained popularity in both industrial and domestic settings because, as compared with conventional composting, it provides a way to treat organic wastes more quickly. In manure composting, it also generates products that have lower salinity levels.

The earthworm species (or composting worms) most often used are red wigglers (*Eisenia fetida* or *Eisenia andrei*), though European nightcrawlers (*Eisenia hortensis* or *Dendrobaena veneta*) could also be used. Red wigglers are recommended by most vermicomposting experts, as they have some of the best appetites and breed very quickly. Users refer to European nightcrawlers by a variety of other names, including *dendrobaenas*, *dendras*, Dutch nightcrawlers, and Belgian nightcrawlers.

Containing water-soluble nutrients, vermicompost is a nutrient-rich organic fertilizer and soil conditioner in a form that is relatively easy for plants to absorb. Worm castings are sometimes used as an organic fertilizer. Because the earthworms grind and uniformly mix minerals in simple forms, plants need only minimal effort to obtain them. The worms' digestive systems create environments that allow certain species of microbes to thrive to help create a "living" soil environment for plants. The fraction of soil which has gone through the digestive tract of earthworms is called the drilosphere.

Design Considerations

Suitable Worm Species

One of the species most often used for composting is the red wiggler or tiger worm (*Eisenia fetida* or *Eisenia andrei*); *Lumbricus rubellus* (a.k.a. red earthworm or dilong (China)) is another breed of worm that can be used, but it does not adapt as well to the shallow compost bin as does *Eisenia fetida*. European nightcrawlers (*Eisenia hortensis*) may also be used. Users refer to European nightcrawlers by a variety of other names, including dendrobaenas, dendras, and nightcrawlers. African Nightcrawlers (*Eudrilus eugeniae*) are another set of popular composters. *Lumbricus terrestris* (a.k.a. Canadian nightcrawlers (US) or common earthworm (UK)) are not recommended, since they burrow deeper than most compost bins can accommodate. Blueworms (*Perionyx excavatus*) may be used in the tropics.

These species commonly are found in organic-rich soils throughout Europe and North America and live in rotting vegetation, compost, and manure piles. They may be an invasive species in some areas. As they are shallow-dwelling and feed on decomposing plant matter in the soil, they adapt easily to living on food or plant waste in the confines of a worm bin.

Composting worms are available to order online, from nursery mail-order suppliers or angling shops where they are sold as bait. They can also be collected from compost and manure piles. These species are not the same worms that are found in ordinary soil or on pavement when the soil is flooded by water.

Large Scale

Large-scale vermicomposting is practiced in Canada, Italy, Japan, India, Malaysia, the Philippines, and the United States. The vermicompost may be used for farming, landscaping, to create compost tea, or for sale. Some of these operations produce worms for bait and/or home vermicomposting.

There are two main methods of large-scale vermiculture. Some systems use a windrow, which consists of bedding materials for the earthworms to live in and acts as a large bin; organic material is added to it. Although the windrow has no physical barriers to prevent worms from escaping, in theory they should not due to an abundance of organic matter for them to feed on. Often windrows are used on a concrete surface to prevent predators from gaining access to the worm population.

The windrow method and compost windrow turners were developed by Fletcher Sims Jr. of the Compost Corporation in Canyon, Texas. The Windrow Composting system is noted as a sustainable, cost-efficient way for farmers to manage dairy waste.

The second type of large-scale vermicomposting system is the raised bed or flow-through system. Here the worms are fed an inch of "worm chow" across the top of the bed, and an inch of castings are harvested from below by pulling a breaker bar across the large mesh screen which forms the base of the bed.

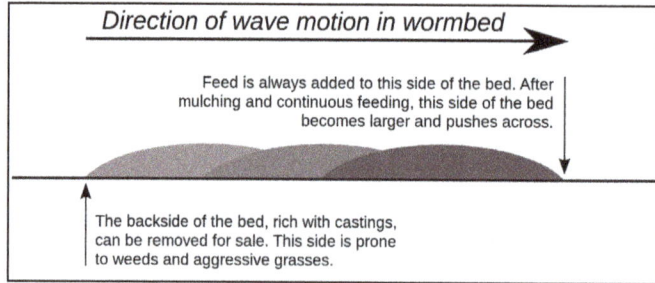

Movement of castings through a worm bed.

Because red worms are surface dwellers constantly moving towards the new food source, the flow-through system eliminates the need to separate worms from the castings before packaging. Flow-through systems are well suited to indoor facilities, making them the preferred choice for operations in colder climates.

Small Scale

Demonstration home scale worm bin at a community garden site (painted plywood).

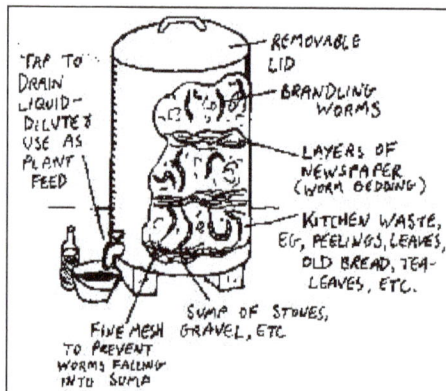

Diagram of a household-scale worm composting bin.

For vermicomposting at home, a large variety of bins are commercially available, or a variety of adapted containers may be used. They may be made of old plastic containers, wood, Styrofoam, or metal containers. The design of a small bin usually depends on where an individual wishes to store the bin and how they wish to feed the worms.

Some materials are less desirable than others in worm bin construction. Metal containers often conduct heat too readily, are prone to rusting, and may release heavy metals into the vermicompost. Styrofoam containers may release chemicals into the organic material. Some cedars, Yellow cedar, and Redwood contain resinous oils that may harm worms, although Western Red Cedar has excellent longevity in composting conditions. Hemlock is another inexpensive and fairly rot-resistant wood species that may be used to build worm bins.

Bins need holes or mesh for aeration. Some people add a spout or holes in the bottom for excess liquid to drain into a tray for collection. The most common materials used are plastic: recycled polyethylene and polypropylene and wood. Worm compost bins made from plastic are ideal, but require more drainage than wooden ones because they are non-absorbent. However, wooden bins will eventually decay and need to be replaced.

Small-scale vermicomposting is well-suited to turn kitchen waste into high-quality soil amendments, where space is limited. Worms can decompose organic matter without the additional human physical effort (turning the bin) that bin composting requires.

Composting worms which are detritivorous (eaters of trash), such as the red wiggler *Eisenia fetidae*, are epigeic (surface dwellers) and together with symbiotic associated microbes are the ideal vectors for decomposing food waste. Common earthworms such as *Lumbricus terrestris* are anecic (deep burrowing) species and hence unsuitable for use in a closed system. Other soil species that contribute include insects, other worms and molds.

Climate and Temperature

There may be differences in vermicomposting method depending on the climate. It is necessary to monitor the temperatures of large-scale bin systems (which can have high heat-retentive properties), as the raw materials or feedstocks used can compost, heating up the worm bins as they decay and killing the worms.

The most common worms used in composting systems, redworms (*Eisenia foetida, Eisenia andrei,* and *Lumbricus rubellus*) feed most rapidly at temperatures of 15–25 °C (59-77 °F). They can survive at 10 °C (50 °F). Temperatures above 30 °C (86 °F) may harm them. This temperature range means that indoor vermicomposting with redworms is possible in all but tropical climates. Other worms like Perionyx excavatus are suitable for warmer climates. If a worm bin is kept outside, it should be placed in a sheltered position away from direct sunlight and insulated against frost in winter.

Feedstock

There are few food wastes that vermicomposting cannot compost, although meat waste and dairy products are likely to putrefy, and in outdoor bins can attract vermin. Green waste should be added in moderation to avoid heating the bin.

Small-scale or Home Systems

Such systems usually use kitchen and garden waste, using "earthworms and other microorganisms to digest organic wastes, such as kitchen scraps". This includes:

- All fruits and vegetables (including citrus, in limited quantities),
- Vegetable and fruit peels and ends,
- Coffee grounds and filters,
- Tea bags (even those with high tannin levels),
- Grains such as bread, cracker and cereal (including moldy and stale),
- Eggshells (rinsed off),
- Leaves and grass clippings (not sprayed with pesticides),
- Newspapers (most inks used in newspapers are not toxic),
- Paper toweling (which has not been used with cleaners or chemicals).

Large-scale or Commercial

Such vermicomposting systems need reliable sources of large quantities of food. Systems presently operating use:

- Dairy cow or pig manure Sewage sludge,
- Brewery waste,
- Cotton mill waste,
- Agricultural waste,
- Food processing and grocery waste,
- Cafeteria waste,
- Grass clippings and wood chips.

Harvesting

Vermicompost is ready for harvest when it contains few-to-no scraps of uneaten food or bedding. There are several methods of harvesting from small-scale systems: "dump and hand sort", "let the worms do the sorting", "alternate containers" and "divide and dump."

These differ on the amount of time and labor involved and whether the vermicomposter wants to save as many worms as possible from being trapped in the harvested compost.

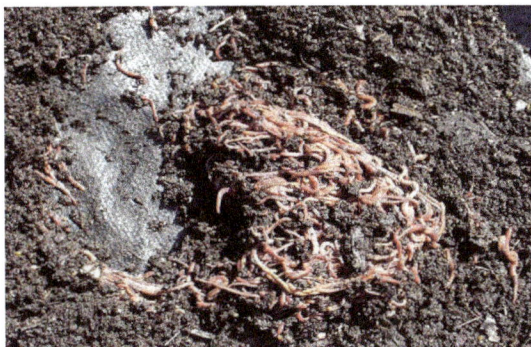

The pyramid method of harvesting worm compost is commonly used in small-scale vermiculture, and is considered the simplest method for single layer bins. In this process, compost is separated into large clumps, which is placed back into composting for further breakdown, and lighter compost, with which the rest of the process continues. This lighter mix is placed into small piles on a tarp under the sunlight. The worms instinctively burrow to the bottom of the pile. After a few minutes, the top of the pyramid is removed repeatedly, until the worms are again visible. This repeats until the mound is composed mostly of worms.

When harvesting the compost, it is possible to separate eggs and cocoons and return them to the bin, thereby ensuring new worms are hatched. Cocoons are small, lemon-shaped yellowish objects that can usually be seen with the naked eye. The cocoons can hold up to 20 worms (though 2-3 is most common). Cocoons can lay dormant for as long as two years if conditions are not conducive for hatching.

Properties

Vermicompost has been shown to be richer in many nutrients than compost produced by other composting methods. It has also outperformed a commercial plant medium with nutrients added, but levels of magnesium required adjustment, as did pH.

However, in one study it has been found that homemade backyard vermicompost was lower in microbial biomass, soil microbial activity, and yield of a species of ryegrass than municipal compost. It is rich in microbial life which converts nutrients already present in the soil into plant-available forms.

Unlike other compost, worm castings also contain worm mucus which helps prevent nutrients from washing away with the first watering and holds moisture better than plain soil.

Increases in the total nitrogen content in vermicompost, an increase in available nitrogen and phosphorus, as well as the increased removal of heavy metals from sludge and soil have been reported. The reduction in the bioavailability of heavy metals has been observed in a number of studies.

Benefits
Soil

- Improves soil aeration;

- Enriches soil with micro-organisms (adding enzymes such as phosphatase and cellulase);

- Microbial activity in worm castings is 10 to 20 times higher than in the soil and organic matter that the worm ingests;

- Attracts deep-burrowing earthworms already present in the soil;

- Improves water holding capacity.

Plant Growth

- Enhances germination, plant growth, and crop yield;

- Improves root growth and structure;

- Enriches soil with micro-organisms (adding plant hormones such as auxins and gibberellic acid).

Economic

- Biowastes conversion reduces waste flow to landfills;

- Elimination of biowastes from the waste stream reduces contamination of other recyclables collected in a single bin (a common problem in communities practicing Single-stream recycling);

- Creates low-skill jobs at local level;

- Low capital investment and relatively simple technologies make vermicomposting practical for less-developed agricultural regions.

Environmental

- Helps to close the "metabolic gap" through recycling waste on-site;

- Large systems often use temperature control and mechanized harvesting, however other equipment is relatively simple and does not wear out quickly;

- Production reduces greenhouse gas emissions such as methane and nitric oxide (produced in landfills or incinerators when not composted).

Uses

Mid-scale worm bin (1 m X 2.5 m up to 1 m deep), freshly refilled with bedding.

Soil Conditioner

Vermicompost can be mixed directly into the soil, or mixed with water to make a liquid fertilizer known as worm tea.

The dark brown waste liquid, or leachate, that drains into the bottom of some vermi-composting systems is not to be confused with worm tea. It is an uncomposted byprod-uct from when water-rich foods break down and may contain pathogens and toxins. It is best discarded or applied back to the bin when added moisture is needed for further processing.

The pH, nutrient, and microbial content of these fertilizers varies upon the inputs fed to worms. Pulverized limestone, or calcium carbonate can be added to the system to raise the pH.

Operation and Maintenance

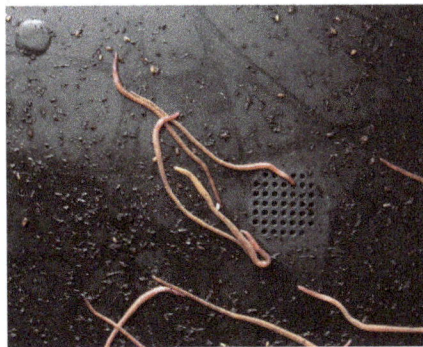

Worms and fruit fly pupas under the lid of a home worm bin.

Smells

When closed, a well-maintained bin is odorless; when opened, it should have little smell—if any smell is present, it is earthy. The smell may also depend on the type of composted material added to the bin. An unhealthy worm bin may smell, potentially due to low oxygen conditions. Worms require gaseous oxygen. Oxygen can be provided by airholes in the bin, occasional stirring of bin contents, and removal of some bin contents if they become too deep or too wet. If decomposition becomes anaerobic from excess wet feedstock added to the bin, or the layers of food waste have become too deep, the bin will begin to smell of ammonia.

Moisture

Moisture must be maintained above 50%, as lower moisture content will not support worm respiration and can increase worm mortality. Operating moisture-content range should be between 70-90%, with a suggested content of 70-80% for vermicomposting-oriented vermiculture operations. If decomposition has become anaerobic, to restore healthy conditions and prevent the worms from dying, excess waste water must be reduced and the bin returned to a normal moisture level. To do this, first reduce addition of food scraps with a high moisture content and second, add fresh, dry bedding such as shredded newspaper to your bin, mixing it in well.

Pest Species

Pests such as rodents and flies are attracted by certain materials and odors, usually from large amounts of kitchen waste, particularly meat. Eliminating the use of meat or dairy product in a worm bin decreases the possibility of pests.

Predatory ants can be a problem in African countries. In warm weather, fruit and vinegar flies breed in the bins if fruit and vegetable waste is not thoroughly covered with bedding. This problem can be avoided by thoroughly covering the waste by at least 5 centimetres (2.0 in) of bedding. Maintaining the correct pH (close to neutral) and water content of the bin (just enough water where squeezed bedding drips a couple of drops) can help avoid these pests as well.

Worms Escaping

Worms generally stay in the bin, but may try to leave the bin when first introduced, or often after a rainstorm when outside humidity is high. Maintaining adequate conditions in the worm bin and putting a light over the bin when first introducing worms should eliminate this problem.

Nutrient Levels

Commercial vermicomposters test, and may amend their products to produce consistent quality and results. Because the small-scale and home systems use a varied mix of feedstocks, the nitrogen, potassium and phosphorus content of the resulting vermicompost will also be inconsistent. NPK testing may be helpful before the vermicompost or tea is applied to the garden.

In order to avoid over-fertilization issues, such as nitrogen burn, vermicompost can be diluted as a tea 50:50 with water, or as a solid can be mixed in 50:50 with potting soil.

Additionally, the mucous layer created by worms which surrounds their castings allows for a "time release" effect, meaning not all nutrients are released at once. This also reduces the risk of burning the plants, as is common with the use and overuse of commercial fertilizers.

Application Examples

Vermicomposting (also known as vermiculture) is widely used in North America for on-site institutional processing of food scraps, such as in hospitals, universities, shopping malls, and correctional facilities. Vermicomposting is used for medium-scale on-site institutional organic material recycling, such as for food scraps from universities and shopping malls. It is selected either as a more environmentally friendly choice than conventional disposal, or to reduce the cost of commercial waste removal.

Researchers discovered that worm composts can also be used to clean up heavy metals. The researchers found substantial reductions in heavy metals when the worms were released into the garbage and they are effective at removing lead, zinc, cadmium, copper and manganese.

Methods of Composting

One of the first choices you need to make is whether continuous or batch composting is best for your needs.

Continuous composting is a technique that works best if you have a steady stream of new material to work with. If you're composting the scraps from your household, this is probably the system you'll want to use. You can start with a small amount of compost and a handful of soil (or compost starter). Then, as you get extra ingredients, just add them to the mix. The compost will blend together — fresh ingredients will blend with more mature compost that's at an advanced stage of decomposition.

As your compost bin starts to fill up, you'll just want to stop adding to it for the last few weeks while you keep mixing up the materials so that the newest materials can finish breaking down too. Alternatively, you can sift out the unfinished materials with a compost screen, and throw them back into the pile or the bin to finish up.

The other method is called batch composting. If you have a large amount of organic waste (such as a pile of leaves or several bags of yard clippings) it can be enough to fill up your entire compost bin all at once. As the compost decomposes, this pile of compost will gradually shrink. Finished compost often takes up about 30 to 50 percent less space space than the original ingredients. It can be tempting to add additional materials to the batch as it starts to shrink and turn into compost, but if you add additional waste, the entire pile of compost will take longer to finish.

Any composter or compost pile can be used for continuous or batch composting or a mix of the two methods. When a continuous composter fills up, it is often converted into a batch composter. As the ingredients compact down, the compost can be left alone (batch composting) or new ingredients can be added as space permits (continuous composting).

Indoor vs. Outdoor Composters

Compost bins come in a wide variety of designs. One of the biggest differences is whether the composter is designed to be used indoors or outdoors. Outdoor compost bins are intended to be placed outside, or in a covered space where odor isn't important (such as a barn or garage). On the other hand, indoor compost bins are designed with odor controls built in. These features tend to either reduce the size of the composter or drive up its price. Indoor compost bins are either air-tight or include a special filter to control smell. The filters have a limited lifespan and need to be replaced every 6 months to a year. They are often made from activated carbon.

Single vs. Multiple Chamber Composters

A single chamber compost bin is the most common. This type of composter has several benefits – large chambers are an optimal size for generating heat, single chamber compost bins have minimal cost, and assembly is usually very simple. On the other hand, single-chamber composters have one major weakness: when the compost bin fills up, there's nowhere left to add waste while the contents decompose. Often, it requires more than one single chamber composter to run staggered batches of compost.

Multiple chamber composters were created to deal with this problem. They offer 2, 3, or more separate compartments for compost. As each compartment fills up, it's possible to seal that compartment and keep adding waste to a different chamber. This allows old waste to completely break down in one part of the composter while new waste is added elsewhere. Uninterrupted composting in the full compartments will quickly yield finished humus, while the additional capacity prevents a backlog of raw waste.

Tumbling vs. Manual Aeration

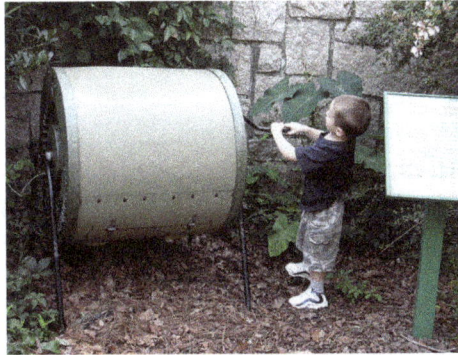

There are many different ways to aerate compost. Most composters are designed with vents that allow air to flow through the bin. Since compost generates heat and hot air expands, composters generate a slight overpressure. As hot air rises from the unit, fresh air is pulled in from below. This airflow doesn't penetrate very far below the surface of the compost though.

These vents are generally not adequate to aerate the entire compost pile. Some composters only have air holes or small vents, which means that a bit of elbow grease is required to aerate their contents. The traditional method is to use a pitchfork. Digging up the compost breaks up any airless pockets and also helps to blend the contents. This is called 'turning' the compost, and it can be hard work. Labor saving compost turning tools are also available, such as compost stirring rods or other specialized compost aerating tools.

Not all compost bins require manual aeration. Self-aerating composters are designed with components that break up clumps and inject fresh air. One of the more popular features is a tumbling bin. Tumbling compost bins are designed with a metal rod through the center that acts as a pivot point. The bins spin around this metal rod, and gravity pulls compost across the metal rod. As the contents of a tumbling composter fall, they break apart against the metal axle and fresh air is blended into the mix. Other composters include fins that stir the compost when it is tumbled or stirring rods that can be cranked with a lever on the side of the bin. Electric and manual models are available.

Compost Bins with Bottoms vs. Ground Sitting Compost Bins

Composters usually sit on the ground, so many composters are designed with only sides and a lid. This reduces the weight of the composter and it also allows compost tea to filter into the earth below. Ground sitting composters can help fight erosion. When bottomless composters are placed over sink holes or cracks in the ground, the finished compost will naturally settle into crevices. Composters without a bottom can even be used to help keep underground bulbs warm during the winter and jump start the germination of certain seeds.

Compost Piles and Trenches

Compost bins make composting easier, but some people prefer to use compost piles or compost trenches. There is no need for a container when composting, but these types of composting require very different ground preparation. Compost piles are built on top of the ground, which should be cleared of weeds or other plants that might intrude. A fully functioning compost pile can get hot enough to sterilize weeds, but piles on the ground rarely reach optimal temperature. They are usually designed to support their own weight, with layers of twigs between other compost. Compost piles occasionally require turning and may need to be watered in dry climates. Compost piles can attract insects, rodents, and larger animals, so they should be located away from buildings and gardens.

Compost trenches, on the other hand, are dug into the ground. Compost pits operate the same way that a latrine or trash midden does, and should be dug at least 3 feet below the surface to keep animals away. Composting is not recommended in areas with bears, and you should check local laws before building a compost pile or digging a trench.

PREPARATION OF MANURE

Organic Manures

Organic manures are composed of dead plant and animal remains and contain plant nutrients. They are applied to the soil to increase crop production. Farmyard manures from cow or buffalo dung, compost made from plants, leaves and kitchen waste, and leguminous crops used as green manures are some examples of organic manures.

Most farmers use compost and other forms of organic manures as supplements to mineral fertilizers. However, compost prepared by traditional methods is not well-decomposed and has a poor nutrient content. Well-decomposed compost will reduce weeds and insects. Also, there are serious insect and weed problems when undecomposed compost is used. The average nitrogen content of the compost prepared by farmers is 0.5%. Using improved methods can increase the nitrogen content of the compost to 1.5%.

Benefits from Compost

- Maintains soil fertility level.

- Increases the nutrient level of the soil or improve the soil's physical condition by improving soil structure and aeration.

- Increases the infiltration capacity of the soil, thus reducing surface runoff.

- Helps to retain plant nutrients and moisture.

- Well-decomposed compost buffers soil reaction and controls soil temperature.

- Increases soil microbial activity which helps mineralization of applied chemical fertilizers, making them more available to crops.

Preparation of Compost

- Arrange composting material in a pit or heap. If composting is done in a heap, the site should be levelled and protected from rain by a roof so that nutrients will not leach.

- Compost is decomposed by fungi and bacteria. For proper microbial growth, add starter materials Complesal (a few handfuls), lime or top soil at each layer. Decomposed compost and wood ash can also be added if chemical fertilizer is not available.

- Add enough water to keep compost moist; the material should be spongy - not too dry, not too wet.

- Turn the compost pit or heap at 30-40 day intervals for proper aeration.

- Cover the compost pit or heap with mudor straw or plastic sheets. This practice enhances decomposition. In the midhills, it may take approximately 3-4 months for complete decomposition.

Watering the compost - Turning the compost.

Well-Decomposed Compost is:

- Friable,

- Does not stick in the hand,

- Dark grey or blackish in color,

- Original material cannot be distinguished.

Using Fresh Banmara for Composting

Eupatorium adenophorum.

Banmara (Eupatorium adenophorum) is a perennial shrub found in the hills (600-2,000 masl). Its vigorous vegetative growth and regeneration take place during the rainy season but remain dormant during the winter. Largely considered as a devastating weed in pastures and forests, this plant can be utilized to prepare compost. Studies reveal that fresh Banmara biomass mixed with cattle dung can produce a good quality compost. The Carbon: Nitrogen (C:N) ratio of the compost prepared from Banmara was 14.2 with 2.0% Nitrogen, 0.02% Phosphorus and 1.2% Potassium.

Preparation

1. Place fresh Banmara biomass (preferably new sprouts; but add a few woody stems to improve aeration) in a 3 ft deep pit in 1-1.5 It layers alternated with thin layers of cattle dung at a ratio of 10:1 (by weight). Use small amounts of other organic materials (lime, wood ash, soil, etc.) and chemical fertilizers (if available) as compost activators.

2. Continue placing alternate layers of Banmara and cattle dung until the materials are about 2-2.5 ft above the soil surface.

3. After 6 weeks, turn the material. One month later, the compost will be ready for use.

If done in the midhills (1,1()()-1,7(N) masl), Barnmara filled in August will only be ready in December, after 5-6 months. Banmara can also be composted in a heap situated in a well-drained spot. Similar materials and alternating layers should be heaped up to four feet above the soil. Other practices associated with the production of good compost should also be practiced.

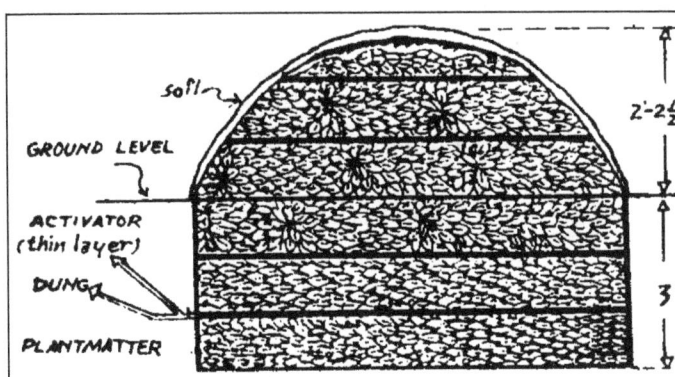

Cross section of compost.

Effective Methods of Compost Application on Hill Farms

Farmers generally carry compost to the field and pile it in small heaps several weeks before ploughing or crop planting, then spread the compost across the field several days before incorporation into the soil. Reports suggest that this practice can result in nitrogen losses of 30-40% due to volatilization. One study indicates that nutrient losses were greater when the compost was spread in the field several days (one week or more) before incorporation, than when simply heaping it in the field. Other disadvantages to this traditional practice include the difficulty of broadcasting compost uniformly throughout the field; the difficulty of incorporating the compost into the soil; and the potential for the compost to wash away in sloping fields.

Maximum use of Compost Nutrients

- Make quality compost using improved methods.

- Protect compost from sun, wind and rain by providing cover or shade.

- If the field where compost is to be used is far from the home or cattle shed, make the compost right in the field. Use large amounts of vegetative material and small amounts of starter/activator materials.

- Always use well-decomposed compost.

- When the compost is ready, store it in a cool, dry place. Pile the compost in a heap and cover with straw or dry soil so that volatilization losses of plant nutrients are minimized.

- Transport the compost when needed. If the compost must remain in the the field for a while, it is better to make only one or two larger heaps in the field rather than many small ones. If possible, cover the heaps with soil.

- When it is time to prepare for crop planting, spread the compost evenly on the land, plough it under and mix thoroughly with the soil. Sow the seeds immediately, unless the compost is not well-decomposed.

- If only limited compost is available and crops are planted with wide spacing, use compost only in small pits where the seed is planted or along the rows/furrows. The farmers' practice of directly sowing potatoes in compost (reducing the amount of compost used by as much as 2/3) is one good example. For grain crops, the compost can be directly spread into the ploughed furrow immediately after ploughing.

- For garden crops, make trenches, place the compost in the trench and cover with soil.

- For tree crops, dig a shallow trench around the tree away from the trunk, just below the drip line. Place the compost in the trench and cover with soil. The same applies for well-established trees as well.

Right – Wrong

Garden and tree crops.

Efficient Method of Organic Matter Application for Maize

Application of organic matter (OM) plays a vital role in the hill darming systems. Farmers know the importance of OM and try to apply maximum amounts in various forms: farm yard manure (FYM), compost, crop residues and green manure. However, most farmers keep OM in small heaps in the field (particularly in upland, i.e., Bari), then broadcast, mix it in the soil and plant the maize after a few days.

Drawbacks to the Existing System

- Loss of nutients in the compost.

- Difficult to broadcast OM uniformly.

- Difficult to incorporate OM in the soil.

- OM away from the maize plants will not be used by the crop, leading to some wastage.

- Erosion losses of OM, along with soil erosion in sloping fields.

The above drawbacks can be corrected if the OM is applied in pits and maize is planted on the sides of the pits. other advantages include: yield increases of more than 30% compared with farmer practices; no need to wait for hired bullocks for land preparation; and increased absorption of fertilizer by plants.

Making Pits

The pits should be about 20 cm deep with a mouth diameter of about 25 cm. The distance between rows should be about 75 cm and within rows, 25 cm. The spacings are approximate and can be adjusted. One ploughing before making the pits will facilitate digging and, thus, will also help control weeds.

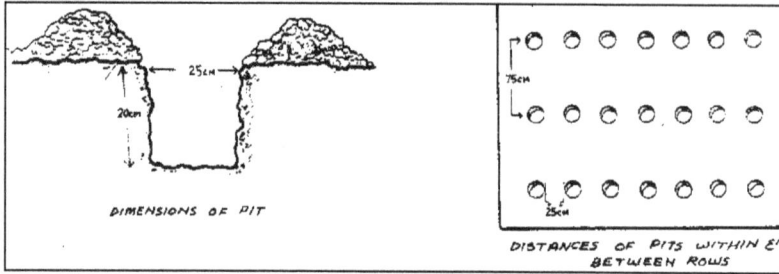

Pits making.

Applying Compost and Fertilizers

Apply about 1150-1250 gm OM per pit. This is equivalent to 30-40 t/ha, which is sufficient for the maize crop. If there is not enough OM, apply whatever is available to the pits and add a mixture of chemical fertilizers (Komplesal+potash) on top of the OM as shown in the figure. This helps to decompose the OM taster and increases the amount of nutrients available to the maize.

Example: If a farmer can only apply 6-# t/ha (230-300 gm/pit), he has to add 7- 10 gm mixture chemical fertilizer per pit. The ratio of OM and fertilizer mixture should be adjusted according to the amount of OM.

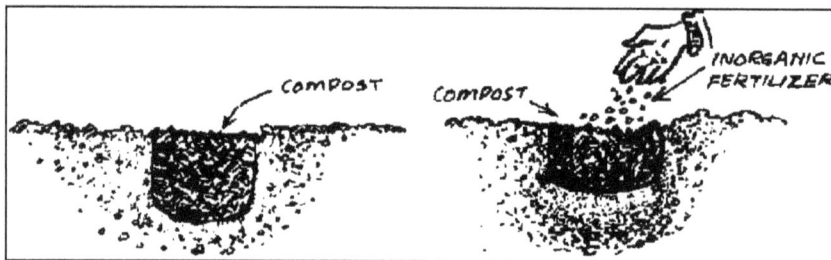

Applying compost and fertilizers.

Planting Maize

After OM and /or chemical application, plant four maize seeds, two on each side of each pit - about 3-5 cm from the edge of the pit. The seeds should be covered with about 5 cm of soil and they should not be in direct contact with the fertilizers.

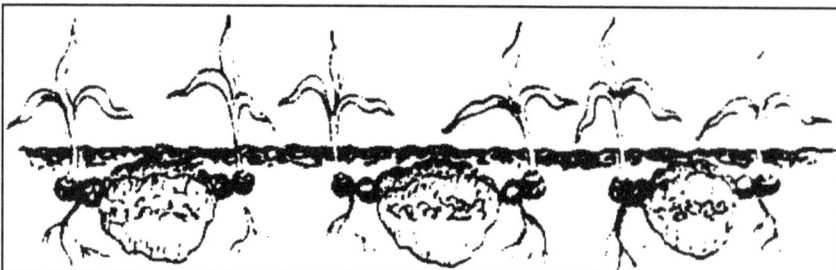

Planting maize.

Thinning

If both seeds germinate, one seedling should be removed, allowing the remaining seedling to grow without competition.

This method is labour-intensive but yield increases are high (up to 39% increase) compared to farmers' traditional practices with manure. Farmers with landlfoldings of less than a half hectare have been interested despite the labour requirement because they need to produce more.

Efficient use of Fertilizer in Maize

Maize is the main food crop. The crop yield depends upon many factors, one of which is plant nutrients. Farmers generally use farm yard manure and compost to supply all the nutrients required. But, due to the increase in cultivation of hybrid and composite varieties of maize, the demand for chemical fertilizers has gone up. However, chemical fertilizers are expensive and are not easily available at the time needed and in the required quantity. Many hill farmers are not able to apply the recommended dose of chemical fertilizer.

In order to get efficient use of minimal amount of chemical fertilizers, the following points should be considered:

1. Apply as much organic manures as possible. Organic manures such as farm yard manure, compost and green manure actually improve the physical condition of th soil and help conserve water, and generally increases the effectiveness of chemical fertilizers.

2. Intercropping with leguminous crops such as soybean, crotalaria, velvet bean and dhaincha should be encouraged when there is adequate moisture. This adds more N to the maize crop.

3. Fertilizer effect is best seen when there is adequate soil moisture. Avoid applying chemical fertilizers when the soil is dry or waterlogged. Drainage should be done under waterloged conditions.

4. Acidic soil should be treated with lime or wood ash to get best use of chemical fertilizers. Organic matter also reduces the negative effects of acidity on corn growth.

5. Remember that weeds also use fertilizers. Therefore, maize must be weeded 30-40 days after planting to avoid wasting fertilizer on growing weeds. Weeds must be controlled in maize fields. If weeds cannot be controlled, then no fertilizers should be used.

6. Apply fertilizer when the plants need it most. Nitrogen is needed most when the plant is about knee high (30 cm) and just before the plant begins tasseling.

7. It possible use hunger signs and deficiency symptoms to find which nutrient are needed most.

8. Soil test, plant tests and results of fertilizer trials can also be used where possible.

Efficient use of fertilizer in maize.

When to Apply

Phosphorus and Potash arc immobile nutrients and therefore, should be applied either at the time of planting or before planting the maize seed. Nitrogen is a mobile nutrient. It should be applied in split doses at the peak requirement period. If nitrogenous fertilizer is in short supply, it should be top dressed when the maize crop is knee high (30 cm) after about a month of sowing and when the crop is tasseling.

How to Apply

In order to get efficient use of minimal fertilizers, it should either be drilled near the plants or band placed/side dressed along the rows of the crop. Broadcasting of fertilizer should be discouraged.

The fertilizer should be covered with soil as soon as it is applied; otherwise more than one half of the nitrogen gets lost by volatalization.

Deficiency Symptoms

Nitrogen deficiency

Nitrogen Deficiency: Pale green or yellowish green appearance of leaves. Stunted growth and reduced size of foliage. A yellowish color extends from the tips along the midribs in a V-shape. Ears are small and kernels not filled.

Phosphorus Deficiency: Purple coloration of leaves. Poor root development and weak stalk. Short and twisted ear with underdeveloped kernels.

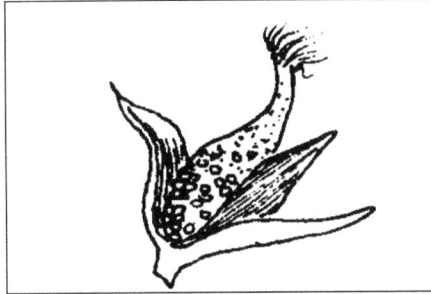

Phosphorus deficiency

Potash Deficiency: Brownillg/brollzillg and scorching of the leaf margins. Dwarf, short internodes. Ears with poorly filled tips.

Potash deficiency

Indigenous Species for Green Manuring

With the introduction of high-yielding varieties (HYV) of cereals and vegetables that fit into multiple cropping systems, there has been a significant reduction in the fertility status of soil in the hills. Also, heavy pressure on forest resources has resulted in a drastic depletion of forest and green vegetation, causing a decrease in the productivity and number of livestock in the hills. Both of these factors contribute to declining soil fertility.

The use of chemical fertilizers is minimal owing to the low purchasing power of the farmers, the high cost of transportation, high chemical fertilizer prices and the unavailability of these fertilizers at the proper time. In this context, it is important to look for locally available, sustainable sources for maintaining soil fertility. Among the locally available sources of maintaining soil fertility, indigenous green manure species occupy an important place.

Chopping and carrying manure.

Green manuring is the process of incorporation of green vegetation, either from cropped land, from marginal land or forest, into the soil for the purpose of improving soil structure and fertility.

More than 20 indigenous plant species having green manure values have been identified and studied. Seven of these species, identified as most important, will be briefly described.

Asuro (Adhatoda vasica): Acantheceae family. Research suggests that the use of Asuro (10 t/ha) increased rice yield by 39% over farmers' traditional practice (10 ton compost,' ha) and 49% over the use of inorganic fertilizers (60:30:30 kg NPK/ha). Planting Asuro as a live fence, can provide a substantial amount of green leaf material. The optimum time for Asuro propagation is from April to mid-June. Asuro is more commonly used in rice and finger millet nurseries and as a mulch in potatoes, transplanted chills and other vegetables. Its leaves and flowers are highly valued as traditional medicine for cough and asthma.

Asuro

Titepati (Artemisia vulgaris): Compositae family. Experiments have shown that the use of Titepati (10 t/ha) in rice enhances vegetative growth and increases rice yield by 23% as compared to inorganic fertilizers (60:30:30 kg NPK/ha). However, increases in rice yield is low as compared to Asuro. Like Asuro it also readily decomposes. Whole plants should be collected, chopped into pieces and buried in the puddled field before transplanting rice. Titepati is also used in rice and finger millet nursery beds and as a live mulch in maize+soybean cropping systems and for garlic cultivation. It has good pesticidal and medicinal values.

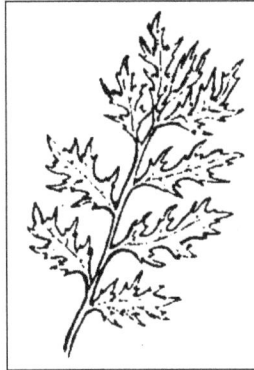

Titepati

Khirro (kapium insigne): Research shows that the use of (10 t/ha) of Khirro as a green manure increased rice yield by 21% over the use of inorganic fertilizers (60:30:30 kg NPK/ha). Khirro leaves decompose easily and can reduce the incidence of rice blast in nursery beds. It is also commonly used to control crab problems. The milky sap of Kh-irro is irritating to human beings.

Rice bean (Vigna unbellata): An important green manure with prolific nodulating potential, Rice bean can be grown as a sole crop in a rice-wheat system and incorporated into the soil after the wheat harvest. Or, it can be grown as a relay crop under spring maize at the second earthing-up operation and ploughed under after the maize harvest to serve as a green manure for normal rice planting. The use of rice bean increases rice yield by 21% over the farmer's practice and is comparable to the use of (60:30:30 kg NPK / ha).

Siris (Albizia lebbeck): A multipurpose leguminous tree species used for fuelwood, timber and as a green manure. The effect of siris leaves on rice yield has been found encouraging; however, its distribution and availability is limited. Use of siris increases rice yield by 34% over farmers' practice and 18% over the use of inorganic fertilizers (60:30:30 kg NPK/ha). Since it does not shade other crops as much as other species, it may be suitable for agroforestry.

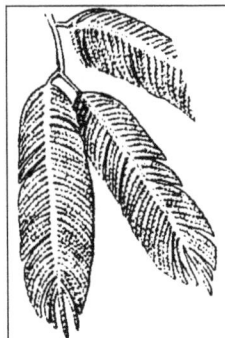

Siris

Padke (Albizia odoretissima): A multipurpose leguminous tree, it is valued as fodder, fuelwood and green manure tree. It is rich in nitrogen and decomposes easily. However, the crop response is inferior to Asuro. Padke is more suitable for mulching potato crop. Padke twigs can be left in the field to drop their leaves and later incorporated into the soil.

Padke

Ankhitare (Walsura trijuga): It is mainly used for rice nurseries and the oil extracted from the seed is valued for medicinal use. It decomposes easily, however its availability is limited.

Other Species with Green Manuring Values

- Bakaino (Melia azaderach)

- Bakula simi (Vicia faba)

- Banmara (Eupatorium adenophorum)

- Bardelo (Ficus clavata)

- Chilaune (Schima wallichii)

- Dhondia (Aeschynomene aspera)

- Jhuse Til (Guizotia abbysinica)

- Kanike phool (Sambucus hooker)

- Pumpkin (Cucurbita moschata)

- Rato siris (Albizia procera).

- Sajion (Jatropa curcas)

- Saljiwan (Origanum vulgare)

Green-Manure Rice Fields using Indigenous Plants

1. Plough the field at least once and irrigate where green manuring plants are to be incorporated.

2. Collect fresh, succulent twigs and leaves of the plant.

3. Chop twigs and leaves into small pieces. Spread the material uniformly over tile field.

4. Plough tile field well to incorporate green biomass into the soil at the rate of IQ t/ha (if available, increase the amount of green biomass up to 20 t/ha). Maintain a shallow water level for 7-10 days.

5. Puddle the soil and transplant rice seedlings.

Table: Comparative information of indigenous green manure species.

Species (Local Name) (Botanical Name)	N (%)	P (%)	K (%)	Elevation Range (m)	Distribution	Plantype	Propogation Method	Remarks
1. Asuro (adhatoda vasica)	4.3	0.9	4.5	Up to 1,300	• Moist or shady areas • Dry, red soils	Dictory ledon-ous, ever-green shrub	Branch cutting (soft and hard wood cutting)	Good for live lence
2. Titepati (Artemisia vulgaris)	2.4	0.9	4.9	Up to 1,2000	• Slopes • Road sides • Fields bunds • Fallow land • Forest areas	Erecl, aeomatic perennial semi-shrub	Seed	Fast-grow-ingand succulent
3. Khirro (supium insigne)	2.8	0.8	2.9	Up to 1,500	• Moist areas • Stream banks • Marginal lands • Forest areas	Tree	Branch cutting	Sap is poi-sonous and cause itivh-ing useful for crab control
4. Rice bean (vigna umbellata)	-	-	-	Sub-trop-ical to mi-hills up to 1,800 m	• From teraj to high hills, dry to wet condi-tions	Legu-minous crop	Seed @ 50kg/ha	Fixes nitrogen
5. Siris (Albizia lebbeck)	2.9	4.5* 0.65	4.5* 2.6	Up to 1,500 m	• River banks • Marginal lands • Forest	Legume tree	Seed	Has tim-ber, fuel wood and agroforest-ry value
6. Padke (Albizia odoretissi-ma)	-	-	-	Up to 800m	• River valleys • Warrner regions	Legume tree	Seed	Valued as a timber and fuel wood, fast growing
7. Ankhitare (walsura trijuga)	2.6	0.49	1.2* 2.4	Up to 1,300m	• Sub-tropical sloping land • River vallevs	Forest tree	Seed	Species nearly dis-appearing

Sesbania Cannabina and Sesbania Rostrata as Green Manures

Recently, the use of organic manures has been overlooked in favour of chemical fertilizers. However, with the continuing world energy crisis and increasing fertilizer prices, green manures are again gaining popularity. Interest in green manures has also been revived because of the increased concern for maintaining long-term soil productivity and ecological sustainability. This interest has led to the identification of lesser known legume plants that have green manuring potential, e.g. Sesbania rostrata, etc.

ADVANTAGES OF MANURE

- These are a good source of macronutrients.

- Improves soil fertility.

- Cost-effective.

- Reduces soil erosion and leaching.

- Improves the physical properties of the soil and aerates the soil.

- Improves the water and nutrient holding capacity of the soil.

- It helps in killing weeds and pests.

- It can be transported easily.

- Methane gas is evolved as the by-product of manure that can be used for cooking and heating purposes.

- The crops grown on the land treated with manure produces healthy crops.

Manure is an ideal soil amendment. When it is applied to the agricultural fields it acts as a field residue. Farmers can sell the manure to people who need to improve their soil fertility. Thus, it can bring income to farmers. They add to the overall soil ability and sustainability. Manure increases the water holding capacity of the soil. The soil organic content can also be improved by applying raw manure like biochar, compost, etc.

Different types of manure contain about 26% solid. The solid and liquid portions are segregated and the solids are used for bedding. The carbon content and other elements can be used to produce different biofuels. Manure also contains a large number of fibres. The undigested animal feed, straw, sawdust, or other bedding contains a lot of fibre.

Manure is environment-friendly and has worked a great deal in increasing food production. It was very difficult to feed a growing population. Use of manure improved the fertility of the soil and increased the yield of the crops.

FERTILIZERS

A fertilizer is any material of natural or synthetic origin (other than liming materials) that is applied to soils or to plant tissues to supply one or more plant nutrients essential to the growth of plants. Many sources of fertilizer exist, both natural and industrially produced.

Mechanism

Fertilizers enhance the growth of plants. This goal is met in two ways, the traditional one being additives that provide nutrients. The second mode by which some fertilizers act is to enhance the effectiveness of the soil by modifying its water retention and aeration. This article, like many on fertilizers, emphasises the nutritional aspect. Fertilizers typically provide, in varying proportions:

- Three main macronutrients:

  - Nitrogen (N): leaf growth.

  - Phosphorus (P): Development of roots, flowers, seeds, fruit.

  - Potassium (K): Strong stem growth, movement of water in plants, promotion of flowering and fruiting.

- Three secondary macronutrients: calcium (Ca), magnesium (Mg), and sulfur (S).

- Micronutrients: copper (Cu), iron (Fe), manganese (Mn), molybdenum (Mo), zinc (Zn), boron (B). Of occasional significance are silicon (Si), cobalt (Co), and vanadium (V).

The nutrients required for healthy plant life are classified according to the elements, but the elements are not used as fertilizers. Instead compounds containing these elements are the basis of fertilizers. The macro-nutrients are consumed in larger quantities and are present in plant tissue in quantities from 0.15% to 6.0% on a dry matter (DM) (0% moisture) basis. Plants are made up of four main elements: hydrogen, oxygen, carbon, and nitrogen. Carbon, hydrogen and oxygen are widely available as water and carbon dioxide. Although nitrogen makes up most of the atmosphere, it is in a form that is unavailable to plants. Nitrogen is the most important fertilizer since nitrogen is present in proteins, DNA and other components (e.g., chlorophyll). To be nutritious to plants, nitrogen must be made available in a "fixed" form. Only some bacteria and their host plants (notably legumes) can fix atmospheric nitrogen (N_2) by converting it to ammonia. Phosphate is required for the production of DNA and ATP, the main energy carrier in cells, as well as certain lipids.

Micronutrients are consumed in smaller quantities and are present in plant tissue on the order of parts-per-million (ppm), ranging from 0.15 to 400 ppm DM, or less than 0.04% DM. These elements are often present at the active sites of enzymes that carry out the plant's metabolism. Because these elements enable catalysts (enzymes) their impact far exceeds their weight percentage.

Classification

Fertilizers are classified in several ways. They are classified according to whether they provide a single nutrient (e.g., K, P, or N), in which case they are classified as "straight fertilizers." "Multinutrient fertilizers" (or "complex fertilizers") provide two or more nutrients, for example N and P. Fertilizers are also sometimes classified as inorganic versus organic. Inorganic fertilizers exclude carbon-containing materials except ureas. Organic fertilizers are usually (recycled) plant- or animal-derived matter. Inorganic are sometimes called synthetic fertilizers since various chemical treatments are required for their manufacture.

Single Nutrient (Straight) Fertilizers

The main nitrogen-based straight fertilizer is ammonia or its solutions. Ammonium nitrate (NH_4NO_3) is also widely used. Urea is another popular source of nitrogen, having the advantage that it is solid and non-explosive, unlike ammonia and ammonium nitrate, respectively. A few percent of the nitrogen fertilizer market (4% in 2007) has been met by calcium ammonium nitrate ($Ca(NO_3)_2 \cdot NH_4 \cdot 10H_2O$).

The main straight phosphate fertilizers are the superphosphates. "Single superphosphate" (SSP) consists of 14–18% P_2O_5, again in the form of $Ca(H_2PO_4)_2$, but also phosphogypsum ($CaSO_4 \cdot 2H_2O$). Triple superphosphate (TSP) typically consists of 44-48% of P_2O_5 and no gypsum. A mixture of single superphosphate and triple superphosphate is called double superphosphate. More than 90% of a typical superphosphate fertilizer is water-soluble.

The main potassium-based straight fertilizer is Muriate of Potash (MOP). Muriate of Potash consists of 95-99% KCl, and is typically available as 0-0-60 or 0-0-62 fertilizer.

Multinutrient Fertilizers

These fertilizers are common. They consist of two or more nutrient components.

Binary (NP, NK, PK) Fertilizers

Major two-component fertilizers provide both nitrogen and phosphorus to the plants. These are called NP fertilizers. The main NP fertilizers are monoammonium phosphate (MAP) and diammonium phosphate (DAP). The active ingredient in MAP is $NH_4H_2PO_4$. The active ingredient in DAP is $(NH_4)_2HPO_4$. About 85% of MAP and DAP fertilizers are soluble in water.

NPK Fertilizers

NPK fertilizers are three-component fertilizers providing nitrogen, phosphorus, and potassium.

NPK rating is a rating system describing the amount of nitrogen, phosphorus, and potassium in a fertilizer. NPK ratings consist of three numbers separated by dashes (e.g., 10-10-10 or 16-4-8) describing the chemical content of fertilizers. The first number represents the percentage of nitrogen in the product; the second number, P_2O_5; the third, K_2O. Fertilizers do not actually contain P_2O_5 or K_2O, but the system is a conventional shorthand for the amount of the phosphorus (P) or potassium (K) in a fertilizer. A 50-pound (23 kg) bag of fertilizer labeled 16-4-8 contains 8 lb (3.6 kg) of nitrogen (16% of the 50 pounds), an amount of phosphorus equivalent to that in 2 pounds of P_2O_5 (4% of 50 pounds), and 4 pounds of K_2O (8% of 50 pounds). Most fertilizers are labeled according to this N-P-K convention, although Australian convention, following an N-P-K-S system, adds a fourth number for sulfur, and uses elemental values for all values including P and K.

Micronutrients

The main micronutrients are molybdenum, zinc, boron, and copper. These elements are provided as water-soluble salts. Iron presents special problems because it converts to insoluble compounds at moderate soil pH and phosphate concentrations. For this reason, iron is often administered as a chelate complex, e.g., the EDTA derivative. The micronutrient needs depend on the plant and the environment. For example, sugar beets appear to require boron, and legumes require cobalt, while environmental conditions such as heat or drought make boron less available for plants.

Production

Nitrogen Fertilizers

Top users of nitrogen-based fertilizer		
Country	Total N use (Mt pa)	Amt. used for feed/pasture (Mt pa)
China	18.7	3.0
India	11.9	N/A

U.S.	9.1	4.7
France	2.5	1.3
Germany	2.0	1.2
Brazil	1.7	0.7
Canada	1.6	0.9
Turkey	1.5	0.3
UK	1.3	0.9
Mexico	1.3	0.3
Spain	1.2	0.5
Argentina	0.4	0.1

Nitrogen fertilizers are made from ammonia (NH_3), which is sometimes injected into the ground directly. The ammonia is produced by the Haber-Bosch process. In this energy-intensive process, natural gas (CH_4) usually supplies the hydrogen, and the nitrogen (N_2) is derived from the air. This ammonia is used as a feedstock for all other nitrogen fertilizers, such as anhydrous ammonium nitrate (NH_4NO_3) and urea ($CO(NH_2)_2$).

Deposits of sodium nitrate ($NaNO_3$) (Chilean saltpeter) are also found in the Atacama desert in Chile and was one of the original nitrogen-rich fertilizers used. It is still mined for fertilizer. Nitrates are also produced from ammonia by the Ostwald process.

Phosphate Fertilizers

All phosphate fertilizers are obtained by extraction from minerals containing the anion PO_4^{3-}. In rare cases, fields are treated with the crushed mineral, but most often more soluble salts are produced by chemical treatment of phosphate minerals. The most popular phosphate-containing minerals are referred to collectively as phosphate rock. The main minerals are fluorapatite $Ca_5(PO_4)_3F$ (CFA) and hydroxyapatite $Ca_5(PO_4)_3OH$. These minerals are converted to water-soluble phosphate salts by treatment with sulfuric (H_2SO_4) or phosphoric acids (H_3PO_4). The large production of sulfuric acid as an industrial chemical is primarily due to its use as cheap acid in processing phosphate rock into phosphate fertilizer. The global primary uses for both sulfur and phosphorus compounds relate to this basic process.

In the nitrophosphate process or Odda process, phosphate rock with up to a 20% phosphorus (P) content is dissolved with nitric acid (HNO_3) to produce a mixture of phosphoric acid (H_3PO_4) and calcium nitrate ($Ca(NO_3)_2$). This mixture can be combined with a potassium fertilizer to produce a *compound fertilizer* with the three macronutrients N, P and K in easily dissolved form.

Potassium Fertilizers

Potash is a mixture of potassium minerals used to make potassium (chemical symbol: K) fertilizers. Potash is soluble in water, so the main effort in producing this nutrient from the ore involves some purification steps; e.g., to remove sodium chloride (NaCl) (common salt). Sometimes potash is referred to as K_2O, as a matter of convenience to those describing the potassium content. In fact, potash fertilizers are usually potassium chloride, potassium sulfate, potassium carbonate, or potassium nitrate.

Compound Fertilizers

Compound fertilizers, which contain N, P, and K, can often be produced by mixing straight fertilizers. In some cases, chemical reactions occur between the two or more components. For example, monoammonium and diammonium phosphates, which provide plants with both N and P, are produced by neutralizing phosphoric acid (from phosphate rock) and ammonia:

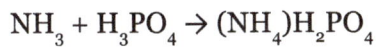

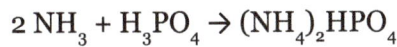

$$NH_3 + H_3PO_4 \rightarrow (NH_4)H_2PO_4$$

$$2\,NH_3 + H_3PO_4 \rightarrow (NH_4)_2HPO_4$$

Organic Fertilizers

Compost bin for small-scale production of organic fertilizer.

A large commercial compost operation.

"Organic fertilizers" can describe those fertilizers with an organic — biologic — origin— that is, fertilizers derived from living or formerly living materials. Organic fertilizers can also describe commercially available and frequently packaged products that strive to follow the expectations and restrictions adopted by "organic agriculture" and "environmentally friendly" gardening — related systems of food and plant production that significantly limit or strictly avoid the use of synthetic fertilizers and pesticides. The "organic fertilizer" *products* typically contain both some organic materials as well as acceptable additives such as nutritive rock powders, ground sea shells (crab, oyster, etc.), other prepared products such as seed meal or kelp, and cultivated microorganisms and derivatives.

Fertilizers of an organic origin (the first definition) include animal wastes, plant wastes from agriculture, compost, and treated sewage sludge (biosolids). Beyond manures, animal sources can include products from the slaughter of animals — bloodmeal, bone meal, feather meal, hides, hoofs, and horns all are typical components. Organically derived materials available to industry such as sewage sludge may not be acceptable components of organic farming and gardening, because of factors ranging from residual contaminants to public perception. On the other hand, marketed "organic fertilizers" may include, and promote, processed organics *because* the materials have consumer appeal. No matter the definition nor composition, most of these products contain less concentrated nutrients, and the nutrients are not as easily quantified. They can offer soil-building advantages as well as be appealing to those who are trying to farm/garden more "naturally".

In terms of volume, peat is the most widely used packaged organic soil amendment. It is an immature form of coal and improves the soil by aeration and absorbing water but confers no nutritional value to the plants. Coir, (derived from coconut husks), bark, and sawdust when added to soil all act similarly (but not identically) to peat and are also considered organic soil amendments - or texturizers - because of their limited nutritive inputs. Some organic additives can have a reverse effect on nutrients — fresh sawdust can consume soil nutrients as it breaks down, and may lower soil pH — but these same organic texturizers (as well as compost, etc.) may increase the availability of nutrients through improved cation exchange, or through increased growth of microorganisms that in turn increase availability of certain plant nutrients. Organic fertilizers such as composts and manures may be distributed locally without going into industry production, making actual consumption more difficult to quantify.

Application

Fertilizers are commonly used for growing all crops, with application rates depending on the soil fertility, usually as measured by a soil test and according to the particular crop. Legumes, for example, fix nitrogen from the atmosphere and generally do not require nitrogen fertilizer.

Liquid vs. Solid

Fertilizers are applied to crops both as solids and as liquid. About 90% of fertilizers are applied as solids. The most widely used solid inorganic fertilizers are urea, diammonium phosphate and potassium chloride. Solid fertilizer is typically granulated or powdered. Often solids are available as prills, a solid globule. Liquid fertilizers comprise anhydrous ammonia, aqueous solutions of ammonia, aqueous solutions of ammonium nitrate or urea. These concentrated products may be diluted with water to form a concentrated liquid fertilizer (e.g., UAN). Advantages of liquid fertilizer are its more rapid effect and easier coverage. The addition of fertilizer to irrigation water is called "fertigation".

Slow and Controlled-release Fertilizers

Slow- and controlled-release involve only 0.15% (562,000 tons) of the fertilizer market. Their utility stems from the fact that fertilizers are subject to antagonistic processes. In addition to their providing the nutrition to plants, excess fertilizers can be poisonous to the same plant. Competitive with the uptake by plants is the degradation or loss of the fertilizer. Microbes degrade many fertilizers, e.g., by immobilization or oxidation. Furthermore, fertilizers are lost by evaporation or leaching. Most slow-release fertilizers are derivatives of urea, a straight fertilizer providing nitrogen. Isobutylidenediurea ("IBDU") and urea-formaldehyde slowly convert in the soil to free urea, which is rapidly uptaken by plants. IBDU is a single compound with the formula $(CH_3)_2CHCH(N-HC(O)NH_2)_2$ whereas the urea-formaldehydes consist of mixtures of the approximate formula $(HOCH_2NHC(O)NH)_nCH_2$.

Besides being more efficient in the utilization of the applied nutrients, slow-release technologies also reduce the impact on the environment and the contamination of the subsurface water. Slow-release fertilizers (various forms including fertilizer spikes, tabs, etc.) which reduce the problem of "burning" the plants due to excess nitrogen. Polymer coating of fertilizer ingredients gives tablets and spikes a 'true time-release' or 'staged nutrient release' (SNR) of fertilizer nutrients.

Controlled release fertilizers are traditional fertilizers encapsulated in a shell that degrades at a specified rate. Sulfur is a typical encapsulation material. Other coated products use thermoplastics (and sometimes ethylene-vinyl acetate and surfactants, etc.) to produce diffusion-controlled release of urea or other fertilizers. "Reactive Layer Coating" can produce thinner, hence cheaper, membrane coatings by applying reactive monomers simultaneously to the soluble particles. "Multicote" is a process applying layers of low-cost fatty acid salts with a paraffin topcoat.

Foliar Application

Foliar fertilizers are applied directly to leaves. The method is almost invariably used to apply water-soluble straight nitrogen fertilizers and used especially for high value crops such as fruits.

Fertilizer burn

Chemicals that affect Nitrogen Uptake

Various chemicals are used to enhance the efficiency of nitrogen-based fertilizers. In this way farmers can limit the polluting effects of nitrogen run-off. Nitrification inhibitors (also known as nitrogen stabilizers) suppress the conversion of ammonia into nitrate, an anion that is more prone to leaching. 1-Carbamoyl-3-methylpyrazole (CMP), dicyandiamide, nitrapyrin (2-chloro-6-trichloromethylpyridine) and 3,4-Dimethylpyrazole phosphate (DMPP) are popular. Urease inhibitors are used to slow the hydrolytic conversion of urea into ammonia, which is prone to evaporation as well as nitrification. The conversion of urea to ammonia catalyzed by enzymes called ureases. A popular inhibitor of ureases is N-(n-butyl)thiophosphoric triamide (NBPT).

Overfertilization

Careful fertilization technologies are important because excess nutrients can be detrimental. Fertilizer burn can occur when too much fertilizer is applied, resulting in damage or even death of the plant. Fertilizers vary in their tendency to burn roughly in accordance with their salt index.

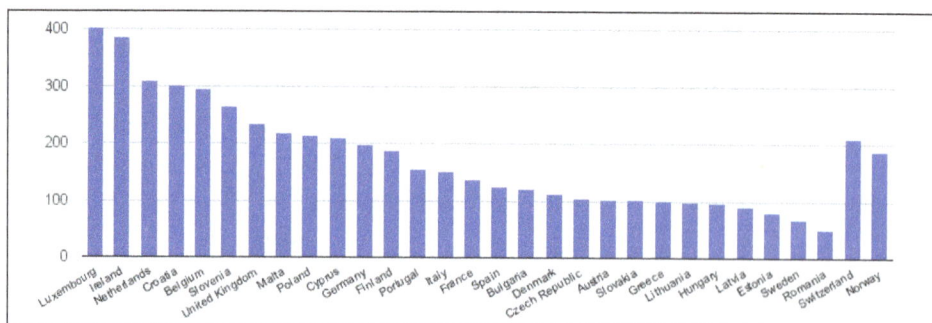

Recently nitrogen fertilizers have plateaued in most developed countries. China although has become the largest producer and consumer of nitrogen fertilizers. Africa has little reliance on nitrogen fertilizers. Agricultural and chemical minerals are very important in industrial use of fertilizers, which is valued at approximately $200 billion. Nitrogen has a significant impact in the global mineral use, followed by potash and phosphate. The production of nitrogen has drastically increased since the 1960s. Phosphate and potash have increased in price since the 1960s, which is larger than the consumer price index. Potash is produced in Canada, Russia and Belarus, together making up over half of the world production. Potash production in Canada rose in 2017 and 2018 by 18.6%. Conservative estimates report 30 to 50% of crop yields are attributed to natural or synthetic commercial fertilizer. Fertilizer consumption has surpassed the amount of farmland in the United States. Global market value is likely to rise to more than US$185 billion until 2019. The European fertilizer market will grow to earn revenues of approximately Euro 15.3 billion in 2018.

For the diagram values of the European Union (EU) countries have been extracted and are presented as kilograms per hectare (pounds per acre). The total consumption of fertilizer in the EU is 15.9 million tons for 105 million hectare arable land area (or 107 million hectare arable land according to another estimate). This figure equates to 151 kg of fertilizers consumed per ha arable land on average for the EU countries.

Environmental Effects

Runoff of soil and fertilizer during a rain storm.

Use of fertilizers are beneficial in providing nutrients to plants although they have some negative environmental effects. The large growing consumption of fertilizers can affect soil, surface water, and groundwater due to dispersion of mineral use.

Water

Phosphorus and nitrogen fertilizers when commonly used have major environmental effects. This is due to high rainfalls causing the fertilizers to be washed into waterways. Agricultural run-off is a major contributor to the eutrophication of fresh water bodies. For example, in the US, about half of all the lakes are eutrophic. The main contributor to eutrophication is phosphate, which is normally a limiting nutrient; high concentrations promote the growth of cyanobacteria and algae, the demise of which consumes oxygen. Cyanobacteria blooms ('algal blooms') can also produce harmful toxins that can accumulate in the food chain, and can be harmful to humans.

The nitrogen-rich compounds found in fertilizer runoff are the primary cause of serious oxygen depletion in many parts of oceans, especially in coastal zones, lakes and rivers. The resulting lack of dissolved oxygen greatly reduces the ability of these areas to sustain oceanic fauna. The number of oceanic dead zones near inhabited coastlines are increasing. As of 2006, the application of nitrogen fertilizer is being increasingly controlled in northwestern Europe and the United States. If eutrophication can be reversed, it may take decades before the accumulated nitrates in groundwater can be broken down by natural processes.

Nitrate Pollution

Only a fraction of the nitrogen-based fertilizers is converted to plant matter. The remainder accumulates in the soil or is lost as run-off. High application rates of nitrogen-containing fertilizers combined with the high water solubility of nitrate leads to increased runoff into surface water as well as leaching into groundwater, thereby causing groundwater pollution. The excessive use of nitrogen-containing fertilizers (be they synthetic or natural) is particularly damaging, as much of the nitrogen that is not taken up by plants is transformed into nitrate which is easily leached.

Nitrate levels above 10 mg/L (10 ppm) in groundwater can cause 'blue baby syndrome' (acquired methemoglobinemia). The nutrients, especially nitrates, in fertilizers can cause problems for natural habitats and for human health if they are washed off soil into watercourses or leached through soil into groundwater.

Soil

Acidification

Nitrogen-containing fertilizers can cause soil acidification when added. This may lead to decrease in nutrient availability which may be offset by liming.

Accumulation of Toxic Elements

Cadmium

The concentration of cadmium in phosphorus-containing fertilizers varies considerably and can be problematic. For example, mono-ammonium phosphate fertilizer may have a cadmium content of as low as 0.14 mg/kg or as high as 50.9 mg/kg. The phosphate rock used in their manufacture can contain as much as 188 mg/kg cadmium (examples are deposits on Nauru and the Christmas islands). Continuous use of high-cadmium fertilizer can contaminate soil (as shown in New Zealand) and plants. Limits to the cadmium content of phosphate fertilizers has been considered by the European

Commission. Producers of phosphorus-containing fertilizers now select phosphate rock based on the cadmium content.

Fluoride

Phosphate rocks contain high levels of fluoride. Consequently, the widespread use of phosphate fertilizers has increased soil fluoride concentrations. It has been found that food contamination from fertilizer is of little concern as plants accumulate little fluoride from the soil; of greater concern is the possibility of fluoride toxicity to livestock that ingest contaminated soils. Also of possible concern are the effects of fluoride on soil microorganisms.

Radioactive Elements

The radioactive content of the fertilizers varies considerably and depends both on their concentrations in the parent mineral and on the fertilizer production process. Uranium-238 concentrations can range from 7 to 100 pCi/g in phosphate rock and from 1 to 67 pCi/g in phosphate fertilizers. Where high annual rates of phosphorus fertilizer are used, this can result in uranium-238 concentrations in soils and drainage waters that are several times greater than are normally present. However, the impact of these increases on the risk to human health from radinuclide contamination of foods is very small (less than 0.05 mSv/y).

Other Metals

Steel industry wastes, recycled into fertilizers for their high levels of zinc (essential to plant growth), wastes can include the following toxic metals: lead arsenic, cadmium, chromium, and nickel. The most common toxic elements in this type of fertilizer are mercury, lead, and arsenic. These potentially harmful impurities can be removed; however, this significantly increases cost. Highly pure fertilizers are widely available and perhaps best known as the highly water-soluble fertilizers containing blue dyes used around households, such as Miracle-Gro. These highly water-soluble fertilizers are used in the plant nursery business and are available in larger packages at significantly less cost than retail quantities. Some inexpensive retail granular garden fertilizers are made with high purity ingredients.

Trace Mineral Depletion

Attention has been addressed to the decreasing concentrations of elements such as iron, zinc, copper and magnesium in many foods over the last 50–60 years. Intensive farming practices, including the use of synthetic fertilizers are frequently suggested as reasons for these declines and organic farming is often suggested as a solution. Although improved crop yields resulting from NPK fertilizers are known to dilute the concentrations of other nutrients in plants, much of the measured decline can be attributed to the use of progressively higher-yielding crop varieties which produce foods with lower mineral concentrations than their less productive ancestors. It is, therefore, unlikely that organic farming or reduced use of fertilizers will solve the problem; foods with high nutrient density are posited to be achieved using older, lower-yielding varieties or the development of new high-yield, nutrient-dense varieties.

Fertilizers are, in fact, more likely to solve trace mineral deficiency problems than cause them: In Western Australia deficiencies of zinc, copper, manganese, iron and molybdenum were identified as limiting the growth of broad-acre crops and pastures in the 1940s and 1950s. Soils in Western Australia are very old, highly weathered and deficient in many of the major nutrients and trace elements. Since this time these trace elements are routinely added to fertilizers used in agriculture in this state. Many other soils around the world are deficient in zinc, leading to deficiency in both plants and humans, and zinc fertilizers are widely used to solve this problem.

Changes in Soil Biology

High levels of fertilizer may cause the breakdown of the symbiotic relationships between plant roots and mycorrhizal fungi.

Energy Consumption and Sustainability

In the US in 2004, 317 billion cubic feet of natural gas were consumed in the industrial production of ammonia, less than 1.5% of total U.S. annual consumption of natural gas. A 2002 report suggested that the production of ammonia consumes about 5% of global natural gas consumption, which is somewhat under 2% of world energy production.

Ammonia is produced from natural gas and air. The cost of natural gas makes up about 90% of the cost of producing ammonia. The increase in price of natural gases over the past decade, along with other factors such as increasing demand, have contributed to an increase in fertilizer price.

Contribution to Climate Change

The greenhouse gases carbon dioxide, methane and nitrous oxide are produced during the manufacture of nitrogen fertilizer. The effects can be combined into an equivalent amount of carbon dioxide. The amount varies according to the efficiency of the process. The figure for the United Kingdom is over 2 kilograms of carbon dioxide equivalent for each kilogram of ammonium nitrate. Nitrogen fertilizer can be converted by soil bacteria to nitrous oxide, a greenhouse gas.

Atmosphere

Through the increasing use of nitrogen fertilizer, which was used at a rate of about 110 million tons (of N) per year in 2012, adding to the already existing amount of reactive nitrogen, nitrous oxide (N_2O) has become the third most important greenhouse gas after carbon dioxide and methane. It has a global warming potential 296 times larger than an equal mass of carbon dioxide and it also contributes to stratospheric ozone depletion. By changing processes and procedures, it is possible to mitigate some, but not all, of these effects on anthropogenic climate change.

Methane emissions from crop fields (notably rice paddy fields) are increased by the application of ammonium-based fertilizers. These emissions contribute to global climate change as methane is a potent greenhouse gas.

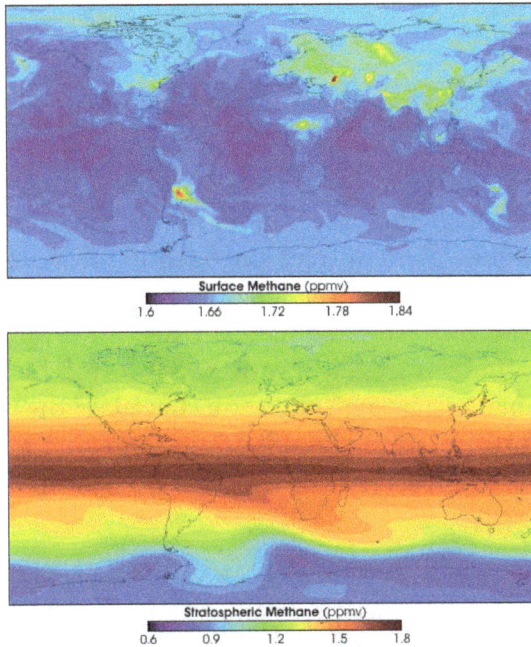

Global methane concentrations (surface and atmospheric) for 2005; note distinct plumes.

Regulation

In Europe problems with high nitrate concentrations in run-off are being addressed by the European Union's Nitrates Directive. Within Britain, farmers are encouraged to manage their land more sustainably in 'catchment-sensitive farming'. In the US, high concentrations of nitrate and phosphorus in runoff and drainage water are classified as non-point source pollutants due to their diffuse origin; this pollution is regulated at state level. Oregon and Washington, both in the United States, have fertilizer registration programs with on-line databases listing chemical analyses of fertilizers.

In China, there have been regulations implemented by the government that want to control N fertilizers being used in farming. In 2008, Chinese governments have started to partially withdraw fertilizer subsidies, which also include contributions to fertilizer transportation, electricity and natural gas use in the industry. Because of this, professional farmers who run large-scale farms have already used less fertilizers since then under the fertilizer prices went up. If large-scale farms keep reducing their use of fertilizer subsidies, they have no choice but to optimize the fertilizer they have which would therefore gain an increase in both grain yield and profit.

Two types of agricultural management practices include organic agriculture and conventional agriculture. The former encourages soil fertility using local resources to maximize efficiency. Organic agriculture avoids synthetic agrochemicals. Conventional agriculture uses all the components that organic agriculture does not use.

NITROGENOUS FERTILIZERS

Nitrogen is absorbed by the plant roots in two forms: nitrate form (NO_3) and ammonical form (NH_4). Most of the crop plants prefer nitrogen in nitrate form; but paddy and few other higher plants prefer nitrogen in ammonical form. Ammonical form of nitrogen is however, easily covertible into nitrate form. Most of the fertilizers contain nitrogen in these two available forms. Urea, however, contains nitrogen in amide form but this form of nitrogen is swiftly converted by soil micro-organisms into ammonical form and then into nitrate form. Based on the forms of nitrogen they contain, nitrogenous fertilizers are classified into following four categories, viz. (1) nitrate fertilizers, (2) ammonical fertilizers, (3) ammonical-nitrate fertilizers, and (4) amide fertilizers.

Nitrate fertilizers: As the name, 'nitrate fertilzers' indicates, these fertilizers contain nitrogen in nitrate form which is preferred by a majority of the plants. These fertilizers are readily soluble in water and quickly available to the plants. If plants need nitrogen urgently, soil use of nitrate fertilizers gives quick response. Use of nitrate fertilizers is not recommended for very moist and waterlogged soils because in these situations nitrate nitrogen is either leached down along with excess moisture or converted into gaseous form and lost to the atmosphere due to lack of oxygen. In both conditions, it becomes unavailable to the plants. In sandy soils also, there use is avoided as the sandy soils are incapable of retaining nutrients and even with a little moisture, nitrate is leached down. If necessary, in sandy soils nitrate fertilizers are applied in 2-3 split doses. There are three main nitrate fertilizers use the world over:

Sodium nitrate	16% Nitrogen (N)
Calcium nitrate	15.5% N
Potassium nitrate	13% N

The use of nitrate fertilizers in our country is very limited due to their limited availability, high cost, difficult handling and also their harmful effect of soil. We have to import these de fertilizers and, so, their cost per unit nutrient (N) is very high. Secondly, when these e-fertilizers are continuously used for a few years, they may deteriorate the physical condition of soil. Plants utilize nitrate portion of these fertilizers and sodium and calcium portions are left in the soil. Excessive build up of sodium or calcium in the soils results in the development of alkalinity or salinity respectively, which is not good for the health of soil.

Ammonical Fertilizers: Ammonical fertilizers contain nitrogen in ammonical form. These fertilizers can be applied also to those crops which absorb nitrogen in nitrate form. It is because ammonical part of the fertilizers can be easily converted into nitrate (NO_3) form. As the ammonical portion of the fertilizer (NH_4) is positively charged, it easily, gets attached with very fine clay or humus a, particles of the soil which are negatively charged. Ammonical fertilizers are, therefore, quite resistant to leaching. These fertilizers can also be applied in waterlogging conditions, because in such conditions, they do not so easily convert into gases. Ammonical fertilizers are especially suitable for paddy crop for a couple a of reasons: paddy prefers nitrogen in ammonical form and ammonical form of nitrogen is not easily converted into gases and lost to atmosphere. Although these fertilizers are readily soluble in water, they are not as quickly available to plants as nitrate fertilizers. A little time is required for the conversion of NH_4 to NO_3, the most preferred form of nitrogen. Ammonical fertilizers are suitable for slow growing, long duration crops. The following ammonical fertilizers are used in our country:

Ammonium sulphate	20% N
Ammonium chloride	24-26% N
Monoammonium phosphate P_2O_5 (Phosphate)	11-20% N, 20-48
Diammonium phosphate (DAP	18%N, 46%P_2O_5
Ammonium solution	20-2S%N

All the ammonical fertilizers exert negative residual effect on the soil. The characteristics of some important ammonical fertilizers can be discussed:

- Ammonium Sulphate: Ammonium sulphate is a whitish, stable crystalline salt, solube in water and stroes well. When applied in the soil, ammonical part gets absorbed on soil clay particles and becomes resistant to leaching. When plants need nitrogen, absorbed ammonium is flushed out into soil solution where micro organisms convert it into nitrate form. Nitrates are then utilised (absorbed) by the plants. The remaining portion of the fertilizer, i.e., sulphate (So_4) is left in the soil and increases acidity in the soil. Application of 100 kg of ammonium sulphate exerts so much of acidity in the soil that it requires 110 kg of calcium carbonate (limestone) to neutralise it. Within a month of its application, ammonium sulphate is utilised by the plants.

- Ammonium chloride: Ammonium chloride is also crystalline ammonical fertilizer available in granular form. It contains 26 per cent nitrogen. This fertilizer is more hygroscopic (absorbing moisture easily from atmosphere) and acidic in residual action than ammonium sulphate. A hundred kg of this fertilizer needs 120 kg neutralizer ($CaCo_3$). The mode of action of this fertilizer is also the same as of ammonium sulphate.

• Diammonium phosphate: Besides nitrogen, diammonium phosphate also contains the second most important primary nutrient element, phosphorous (46% P_2O_5) It is completely solublee in water. It has a very high a content of plant food per bag and has good storage properties. This fertilizer is mainly I considered for supplying phosphate (P_2O_5). The amount of nitrogen virtually supplied through this n fertilizer is then deducted from the total nitrogen requirement of the crop and the rest is applied - through another straight nitrogenous fertilizer.

• Ammonical-nitrate fertilizers: Ammonium-nitrate fertilizers contain nitrogen in both the forms: ammonical and nitrate. Fertilizers of this category can be used in a wide variety of soils and cropping conditions. Nitrate meets the immediate demand of nitrogen and ammonical form supplies nitrogen to meet the growth and developmental needs of N as the plant ages. Fertilizers of this category are acidic in reaction. The following fertilizers of this category are marketed:

Ammonium nitrate	33%N
Calcium ammonium	26%N
Ammonium sulphate Nitrate (ASN)	26%N

Some salient characteristics of these fertilizers are as follows:

• Ammonium nitrate: Although ammonium nitrate is nutritionally very rich, it has several undesirable features. It is highly hygroscopic and shows a marked tendency to absorb moisture from the air and to cake (lump formation). It is highly explosive and, therefore, requires careful handling and storage. This fertilizer is not manufactured in India but imported on a limited scale. The following recommendations can be made to minimize the explosive hazards of this fertilizer:

  ◦ Do not store any longer than is absolutely necessary.

  ◦ Keep all bags tightly sealed until ready to use.

  ◦ Keep all bags indoor in a dry place.

  ◦ Pile the stacks not higher than 6 bags.

  ◦ Change position of bags once a week.

If caking occurs, drop bags on a hard floor.

Some factories have started using conditioners such as clay and one per cent water repellent material like wax, grease or oil to improve the quality of fertilizer. Such granulated material are stored in moisture proof polythene bags.

- Ammonium sulphate nitrate: Ammonium sulphate nitrate fertilizer has the advantage of having nitrogen in both the forms of nitrate and ammonical and also sulphur. It has, however, one disadvantage in that when it is stored it sets into a hard cake which requires to in be broken up. This fertilizer is resistant to leaching as 3/4th of its total nitrogen is in be ammonical form. The acidity released by 100 kg an of this fertilizer requires 93 kg of calcium (a carbonate $CaCO_3$ to neutralize it).

- Calcium ammonium nitrate: Calcium ammonium nitrate fertilizer has very good physical condition. It is attractive, dust free; easy flowing and granular in form. This is, in fact, a chemical mixture of ammonium nitrate with calcium carbonate. Calcium improves the effectiveness of fertilizer due to the following reasons.

  ○ The explosiveness and hygroscopic tendency -of ammonium nitrate is reduced.

  ○ Fertilizer become neutral in reaction leaves neither acidity nor alkalinity as residual effect on soil and can be used in any type of soil.

  ○ Granule size becomes bigger enabling easy application of fertilizer to the field.

  ○ Soil is enriched with calcium, a secondary essential nutrient.

- Amide fertilizers: Amide group fertilizers are agronomically very significant. They are readily soluble in water and easily decomposed by micro organisms in the soil. In the soil, they are quickly changed into ammonical and then nitrate form and becomes available to the plants. The two main amide fertilizers used in our country are: (a) Urea -46%, (b) Calcium cynamid -21%N.

- Urea: Urea is readily soluble in water and is, perhaps, the cheapest among all the nitrogenous fertilziers based on per unit cost of nutrient. It is a white, granular, solid and slightly hygroscopic fertilizer. It can also be applied in solution form as spray. Urea exerts acidity equivalent to neutralizing amount of 80 kg calcium carbonate per 100 kg of fertilizer material. Most of the crops respond effectively to the urea application. When applied to the soil, urea - undergoes decomposition in the presence of enzyme urease', secreted by soil microorganisms. Conversion of urea in to nitrate form due to decomposition is completed in about - one week's time. Urea is somewhat resistant to leaching.

- Calcium cynamide: This fertilizer creates a high basicity/salinity in the & soil as residual effect which requires sulphur as neutralizer. This fertilizer, if available, can be used in acidic soils.

PHOSPHATE FERTILIZERS

Phosphorus plays a key role in the transfer of energy inside the plant. It is essential for photosynthesis (process in which green leaves manufacture carbohydrates in the presence of sunlight) and other physiological processes in the plant. It is indispensable for development of plant body. Plants need it in good amounts, which the soils can not supply without the soils being supplemented from outside. A serious drawback existing with the phosphatic fertilizers is that their phosphorus easily gets fixed with soil clays soon after the fertilizer is applied to the soil. Plants are hardly able to absorb more than 15-20 per cent of the total added fertilizer-phosphorus. The remaining phosphorus is fixed in the soil and becomes unavailable to the plant grant roots. For the phosphorus to be available to the sold plants, it must be soluble in soil solution (soil or p water containing some salts dissolved in it). Solubility of phosphorus in the soil solution depends on the reaction or pH of the soil. It also depends on the type of phosphatic fertilizer. Depending on the olubility of are phosphatic fertilizers, they are grouped into the three categories, viz., water soluble phosphatic fertilizers, citric acid soluble - phosphatic fertilizers, and insoluble phosphatic fertilizers.

POTASSIC FERTILIZERS

Potash is the third most important primary nutrient element required by the plants. Although crop-harvest removes more potash than nitrogen and phosphorus, the soil replenishes it faster than the other two. The deficiency of potash in soils is, therefore, not so marked as of nitrogen and phosphorus. Deficiency of potash in the field is symptomised by scorching of leaf tips advancing towards leaf margins and premature death of leaves even though enough nitrogen and phosphorus are supplied. Increased incidence of insects and pests attack and subsequent lodging of crop plants as the plants loose their mechanical strength are also caused by potash deficiency. Fertilizers supplying potash, the chief commercial ones are the potassium sulphate (50% K2o) and the muriate of potash (60% K2o). The others are relatively of very little importance.

On the basis of the percentage nutrient content, it is better to choose muriate of potash than potassium sulphate. Muriate of potash is also more suitable than potassium sulphate for acidic soils. Potassium sulphate should be preferred to muriate of potash for well aerated, calcareous (too much presence of calcium) and alkaline soils. In highly leached acid soils, where sulphur deficiency is suspected, the use of potassium sulphate can be justified. In both the fertilizers, potassium is utilised by the plants and the remaining portions are left in the soil. If muriate of potash is added to alkaline soil, accumulation of chloride may prove toxic to crops. Therefore, for alkaline soils, potassium sulphate should be preferred. Some crops are also sensitive to chlorlde damage (e.g., potato, tobacco etc). In such crops, potassium sulphate should be used.

COMPOUND FERTILIZERS

Many soils require adding several essential nutrients to alleviate plant deficiencies. Farmers may opt to select a combination of single-nutrient fertilizers or apply a fertilizer that combines several nutrients into each particle. These combination fertilizers (compound or complex) can offer advantages of convenience in the field, economic savings and ease in meeting crop nutritional needs.

Production

Manufacturers make compound fertilizers by using basic fertilizer materials, such as ammonia (NH_3), ammonium phosphate, urea, sulfur (S) and potassium (K) salts. There are many methods used for making these fertilizers, with the specific manufacturing processes determined by the available basic components and the desired nutrient content of the finished product. Here are four brief examples:

- Compaction methods (agglomeration) involve binding small fertilizer particles together using compaction, a cementing agent or a chemical bond. Various nutrient ratios can be combined using undersized particles that may not be suitable for other applications.

- Accretion-based fertilizers are made by repeatedly adding a thin film of nutrient slurry, which is continually dried, building up multiple layers until the desired granule size is reached.

- Pipe-cross reactors are used to chemically melt NH_3, acids containing S or phosphorus (P), and other nutrients—such as K sources and micronutrients—into a solid fertilizer with the desired nutrient content.

- The nitrophosphate process involves reacting phosphate rock with nitric acid to form a mixture of compounds containing N and P. If a K source is added during the process, a solid fertilizer with N, P and K will result.

Agricultural Use

Compound fertilizers contain multiple nutrients in each individual granule. They differ from a blend of fertilizers mixed together to achieve a desired average nutrient composition. This difference allows farmers to spread compound fertilizer so that each granule delivers a mixture of nutrients as it dissolves in the soil and eliminates the potential for segregation of nutrient sources during transport or application. The multiple-nutrient granules of compound fertilizers also allow the farmer to achieve uniform distribution of micronutrients throughout the root zone.

These fertilizers are especially effective for applying an initial nutrient dose in advance of planting. There are certain ratios of nutrients available from a fertilizer dealer for specific soil and crop conditions. This approach offers advantages of simplicity in making complicated fertilizer decisions, but does not allow farmers the flexibility to blend fertilizers to meet specific crop requirements. Turf managers and homeowners often find compound fertilizers especially desirable.

Management Practices

Compound fertilizers are sometimes more expensive than a physical combination or blend of the primary nutrient sources, since they require additional processing. However, when purchasers consider all the factors involved with nutrient handling and use, compound fertilizers may offer considerable advantages.

Among the nutrients, N typically requires the most careful management and reapplication during the growing season. It may not be feasible to supply sufficient N in advance of planting to meet the entire demand (using only compound fertilizer) without overapplying some of the other nutrients. Because of this, growers should consider applying a compound fertilizer early in the growing season and then later add N as needed.

Manufacturers often produce compound fertilizers regionally to meet local crop needs. They typically adjust a wide range of chemical and physical properties to meet those needs. For example, a desire to minimize P in urban storm water runoff has led some communities to restrict the addition of P to compound fertilizers sold for turf and ornamental purposes. In another example, manufacturers customize their products by boosting certain fertilizer elements for regional soils known for deficiency in those nutrients.

NPK FERTILIZER

NPK fertilizer is a complex fertilizer comprised primarily of the three primary nutrients required for healthy plant growth. The agriculture industry relies heavily on the use of NPK fertilizer to meet global food supply and ensure healthy crops.

According to the IFDC, about half of the global population is alive as a result of the increased food production provided by the use of mineral fertilizers.

Components

There are numerous building blocks of life that plants need for healthy and optimum growth. Without these nutrients, plants cannot grow to their full potential, will provide lower yields, and be more susceptible to disease.

The three most important nutrients, without any one of which plants could not survive, are referred to as the primary macronutrients: Nitrogen (N), Phosphorus (P), and Potassium (K).

Soils often lack these nutrients, either naturally, or as a result of over cultivation or other environmental factors. In cases where soils are lacking, nutrients must be put back into the soil in order to create the ideal environment for optimal plant growth.

Each of the primary nutrients is essential in plant nutrition, serving a critical role in the growth, development, and reproduction of the plant.

Nitrogen (N)

Role of Nitrogen in Plants

Nitrogen is a key component in many of the processes needed to carry out growth. In particular, nitrogen is vital to chlorophyll, which allows plants to carry out photosynthesis (the process by which they take in sunlight to produce sugars from carbon dioxide and water). Nitrogen is also a significant component in amino acids, the basis of proteins. Nitrogen also aids in the compounds that allow for storage and use of energy.

One study looked at US cereal yields and how they were affected by omitting nitrogen fertilizer. The study estimated that without nitrogen, average yields for corn declined by a staggering 41%, rice by 37%, barley by 19%, and wheat by 16%.

Sources of Nitrogen

While nitrogen can be taken in and converted into a usable nutrient from the atmosphere, and may be naturally present in soils, it is almost always desirable to supplement nitrogen to ensure plants have the optimum amount available to them. The following materials can be included in NPK blends as a source of nitrogen:

Common Inorganic Sources of N in NPK Blends

- Urea
- Urea Ammonium Nitrate
- Anyhdrous Ammonia

Common Organic Sources of N in NPK Blends

- Manure

- Compost

- Blood Meal

- Feather Meal

Phosphorus (P)

Phosphorus also plays a role in an array of functions necessary for healthy plant growth, contributing to structural strength, crop quality, seed production, and more. Phosphorus also encourages the growth of roots, promotes blooming, and is essential in DNA.

The transformation of solar energy into usable compounds is also largely possible because of phosphorus.

Sources of Phosphorus

Like nitrogen, phosphorus in NPK fertilizer can come from both organic and inorganic sources.

Common Inorganic Sources of P in NPK Blends

The primary source of inorganic phosphorus is phosphate rock. Crushed phosphate rock can be applied to soils directly, but it is much more effective if processed to be more readily available for plant uptake.

Common Organic Sources of P in NPK Blends

- Manure

- Compost

- Biosolids

- Blood Meal

- Bone Meal

Potassium (K)

Potassium is also vital in a variety of other processes that contribute to growth and development. Potassium is often referred to as the "quality element," because of its contribution to many of the characteristics we associate with quality, such as size, shape, color, and even taste, among others.

Plants low in potassium are stunted in growth and provide lower yields.

Sources of Potassium

Potassium can be obtained from a wide range of sources, both organic and inorganic.

Common Inorganic Sources of K in NPK Blends

The primary inorganic source of potassium for use in NPK fertilizers is potash. Like phosphate rock, potash is mined all over the world and processed into a more refined product. Potassium can also come from potassium sulfate, langbeinite, and granite dust.

Common Organic Sources of K in NPK Blends

- Manure

- Compost

- Wood Ash

NPK Fertilizer Production

NPK fertilizer is available in liquid, gaseous, and granular form, with granular being the most common.

Many methods exist for producing a granular NPK fertilizer. Individual components may be produced separately and blended together in specific formulations to create target nutrient ratios, or grades. Or, all-in-one granules containing the desired ratio in each granule may also be produced. The most common approaches to producing granular NPK fertilizer include:

- Pipe Reactor Granulation Systems

- Drum Granulation Systems

- Mixer-Dryer Granulation (Incorporating a pugmill mixer) Systems

- Disc Pelletizing Systems

- Spherodizer Granulation Systems

- Prilling Systems

While NPK fertilizers are comprised mainly of the three primary nutrients, flexibility in processing allows various other micronutrients to be incorporated into the blend. For example, NPKS has been gaining popularity in response to the sulfur deficient soils resulting from the Acid Rain Act.

Specialty fertilizers, or fertilizers formulated to suit the unique nutrient needs of a particular location, are also becoming more popular.

BIOFERTILIZERS

A biofertilizer (also bio-fertilizer) is a substance which contains living microorganisms which, when applied to seeds, plant surfaces, or soil, colonize the rhizosphere or the interior of the plant and promotes growth by increasing the supply or availability of primary nutrients to the host plant. Biofertilizers add nutrients through the natural processes of nitrogen fixation, solubilizing phosphorus, and stimulating plant growth through the synthesis of growth-promoting substances. Biofertilizers can be expected to reduce the use of synthetic fertilizers and pesticides. The microorganisms in biofertilizers restore the soil's natural nutrient cycle and build soil organic matter. Through the use of biofertilizers, healthy plants can be grown, while enhancing the sustainability and the health of the soil. Since they play several roles, a preferred scientific term for such beneficial bacteria is "plant-growth promoting rhizobacteria" (PGPR). Therefore, they are extremely advantageous in enriching soil fertility and fulfilling plant nutrient requirements by supplying the organic nutrients through microorganism and their by-products. Hence, biofertilizers do not contain any chemicals which are harmful to the living soil.

Biofertilizers provide "eco-friendly" organic agro-input. Biofertilizers such as *Rhizobium*, *Azotobacter*, *Azospirilium* and blue green algae (BGA) have been in use a long time. *Rhizobium* inoculant is used for leguminous crops. *Azotobacter* can be used with crops like wheat, maize, mustard, cotton, potato and other vegetable crops. *Azospirillum* inoculations are recommended mainly for sorghum, millets, maize, sugarcane and wheat. Blue green algae belonging to a general cyanobacteria genus, *Nostoc* or *Anabaena* or *Tolypothrix* or *Aulosira*, fix atmospheric nitrogen and are used as inoculations for paddy crop grown both under upland and low-land conditions. *Anabaena* in association with water fern *Azolla* contributes nitrogen up to 60 kg/ha/season and also enriches soils with organic matter.

Other types of bacteria, so-called phosphate-solubilizing bacteria, such as *Pantoea agglomerans* strain P5 or *Pseudomonas putida* strain P13, are able to solubilize the insoluble phosphate from organic and inorganic phosphate sources. In fact, due to immobilization of phosphate by mineral ions such as Fe, Al and Ca or organic acids, the rate of available phosphate (P_i) in soil is well below plant needs. In addition, chemical P_i fertilizers are also immobilized in the soil, immediately, so that less than 20 percent of added fertilizer is absorbed by plants. Therefore, reduction in P_i resources, on one hand, and environmental pollutions resulting from both production and applications of chemical P_i fertilizer, on the other hand, have already demanded the use of phosphate-solubilizing bacteria or phosphate biofertilizers.

Benefits

These are means of fixing the nutrient availability in the soil. Since a bio-fertilizer is technically living, it can symbiotically associate with plant roots. Involved microorganisms could readily and safely convert complex organic material into simple compounds, so that they are easily taken up by the plants. Microorganism function is in long duration, causing improvement of the soil fertility. It maintains the natural habitat of the soil. It increases crop yield by 20-30%, replaces chemical nitrogen and phosphorus by 30%, and stimulates plant growth. It can also provide protection against drought and some soil-borne diseases.

Some important groups of biofertilizers include:

- *Azolla-Anabena* symbiosis: Azolla is a small, eukaryotic, aquatic fern having global distribution.Prokaryotic blue green algae Anabena azolla resides in its leaves as a symbiont. Azolla is an alternative nitrogen source. This association has gained wide interest because of its potential use as an alternative to chemical fertilizers.

- *Rhizobium*: Symbiotic nitrogen fixation by Rhizobium with legumes contribute substantially to total nitrogen fixation. Rhizobium inoculation is a well-known agronomic practice to ensure adequate nitrogen.

ORGANIC FERTILIZER

Organic fertilizers are fertilizers derived from animal matter, animal excreta (manure), human excreta, and vegetable matter (e.g. compost and crop residues). Naturally occurring organic fertilizers include animal wastes from meat processing, peat, manure, slurry, and guano.

A cement reservoir containing cow manure mixed with water. This is common in rural Hainan Province, China. Note the bucket on a stick that the farmer uses to apply the mixture.

In contrast, the majority of fertilizers used in commercial farming are extracted from minerals (e.g., phosphate rock) or produced industrially (e.g., ammonia). Organic agriculture, a system of farming, allows for certain fertilizers and amendments and disallows others.

Compost bin for small-scale production of organic fertilizer.

A large commercial compost operation.

The main organic fertilizers are, peat, animal wastes (often from slaughter houses), plant wastes from agriculture, and treated sewage sludge.

Mineral

By many definitions, minerals are separate from organic materials. However, certain organic fertilizers and amendments are mined, specifically guano and peat. Other mined minerals are fossil products of animal activity, such as greensand (anaerobic marine deposits), some limestones (fossil shell deposits), and some rock phosphates (fossil guano).

Peat is the most widely used organic amendment.

Peat, a precursor to coal, offers no nutritional value to the plants, but improves the soil by aeration and absorbing water. It is sometimes credited as being the most widely use organic fertilizer and by volume is the top organic amendment.

Animal Sources

Animal sourced materials include both animal manures and residues from the slaughter of animals. Manures are derived from milk-producing dairy animals, egg-producing poultry, and animals raised for meat and hide production. When any animal is butchered, only about 40% to 60% of the live animal is converted to market product, with the remaining 40% to 60% classed as by-products. These by-products of animal slaughter, mostly inedible blood, bone, feathers, hides, hoofs, horns, can be refined into agricultural fertilizers including bloodmeal, bone meal fish meal, and feather meal.

Chicken litter, which consists of chicken manure mixed with sawdust, is an organic fertilizer that has been proposed to be superior for conditioning soil for harvest than synthetic fertilizers.

Plant

Processed organic fertilizers include compost, humic acid, amino acids, and seaweed extracts. Other examples are natural enzyme-digested proteins. Decomposing crop residue (green manure) from prior years is another source of fertility.

Other ARS studies have found that algae used to capture nitrogen and phosphorus runoff from agricultural fields can not only prevent water contamination of these nutrients, but also can be used as an organic fertilizer. ARS scientists originally developed the "algal turf scrubber" to reduce nutrient runoff and increase quality of water flowing into streams, rivers, and lakes. They found that this nutrient-rich algae, once dried, can be applied to cucumber and corn seedlings and result in growth comparable to that seen using synthetic fertilizers.

Treated Sewage Sludge

Sewage sludge, also known as biosolids, is effuent that has been treated, blended, composted, and sometimes dried until deemed biologically safe. As a fertilizer it is most commonly used on non-agricultural crops such as in silviculture or in soil remediation. Use of biosolids in agricultural production is less common, and the National Organic Program of the USDA (NOP) has ruled that biosolids are not permitted in organic food production in the U.S.; while biologic in origin (vs mineral), sludge is unacceptable due to toxic metal accumulation, pharmaceuticals, hormones, and other factors.

With concerns about human borne pathogens coupled with a growing preference for flush toilets and centralized sewage treatment, biosolids have been replacing night soil (from human excreta), a traditional organic fertilizer that is minimally processed.

UREA FERTILIZER

The urea fertilizer, also popularly called forty six zero zero (46-0-0), is a simple or straight (single-element) fertilizer that supplies the major essential element nitrogen in ammonic form (NH_4^+). The positively charged ammonium ion (NH_4^+) is nonvolatile and is one of the two forms of nitrogen that can be absorbed by plants, the other being nitrate (NO_3^-).

Urea is the richest source of nitrogen among the common dry fertilizers. Anhydrous ammonia (NH_3), which contains 82% nitrogen, is a pressurized liquid that transforms into gas when released (liquified gas). According to Jones et al., urea ranks as the most preferred dry nitrogenous fertilizer in the United States due to advantages such as high nutrient analysis, easy handling, and reasonable price per unit of nitrogen.

Urea is the most widely used fertilizer worlwide. In 1961 the world consumption of urea was only 2 million tons as against 12 million tons for ammonium sulfate (21-0-0). But the former has continued to grow and, as of 1998, has surpassed the latter by a large margin (86.13 against 12.5 million tons).

According to FAO, there would be a steady rise in the world demand for the fertilizer nutrient nitrogen from 105.3 million tons to 112.9 million tons at the rate of 1.7% per annum. This forecast is consistent with the predicted increase in the production of major food crops which is necessitated by a continuing population surge. Consequently, 58 more urea plants will become operational, seventeen of which are to be located in China.

Urea Fertilizer: 46-0-0

The alternative name 46-0-0 of urea fertilizer stands for its NPK content (actually N, P_2O_5 and K_2O or nitrogen, phosphate and potash, respectively). It means that it contains 46% nitrogen (N), zero phosphorus, and zero potassium. Thus 100 kg of granular urea supplies 46 kg N with the remainder consisting of carriers or fillers.

Granular urea supplies about 46% nitrogen in ammonic form.

46% N as shown on packaging is a nutritional content of the fertilizer that is warranted by the manufacturer. However, it is not exactly the same as the computed value:

Chemical formula of urea = $CO(NH_2)_2$ = CH_4N_2O

- C_1 = 12.01 x 1 = 12.01;

- H_4 = 1.01 x 4 = 4.04;

- N_2 = 14.01 x 2 = 28.02;

- O_1 = 16.0 x 1 = 16.0;

- Total = 12.01 + 4.04 + 28.02 + 16.0 = 60.07;

- Percent nitrogen = 28.02/60.07 x 100 = 46.6%.

Applying Urea Fertilizer

Urea can be applied in the soil in the form of solid granules or prills, or pellets. The most dominant formulation is the white, crystalline granule. The prills used to be the primary form of urea, but they have been surpassed by granules which are larger, harder, and more stable under high humidity. Urea can also be dissolved in water and used as a soil drench or otherwise distributed with irrigation water or applied as foliar spray.

When urea fertilizer is applied to the soil, it combines with water (hydrolysis) to form ammonium carbonate [$(NH_4)_2CO_3$] through the catalytic action of urease. The enzyme urease is present in the soil, resulting from the decomposition of organic matter by microorganisms.

Ammonium carbonate is unstable. It decomposes into gaseous ammonia (NH_3), carbon dioxide, and water. When incorporated to the soil, NH_3 is converted to ammonium (NH_4^+) with hydrogen ion (H^+) coming from soil solution or from soil particles. The positively charged ammonium ions are then fixed into the negatively charged soil particles where they remain until absorbed by plant through the roots or used by bacteria as source of energy and converted to nitrate (NO_3^-) in the process of nitrification.

A summary reaction for the hydrolysis of urea [$CO(NH_2)_2$] leading to the formation of ammonium ion (NH_4^+), a form of N the plants can absorb, is:

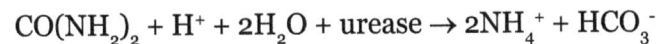

$$CO(NH_2)_2 + H^+ + 2H_2O + urease \rightarrow 2NH_4^+ + HCO_3^-$$

While applying Urea Fertilizer:

- Apply urea by soil incorporation. As a general rule, urea should not be applied on the soil surface or top dressed on sods or crop residues without immediate incorporation. When applied on the soil surface, NH_3, a product of urea hydrolysis, will escape into the air being a gas. This is called ammonia volatilization.

- Volatilization loss from nitrogenous fertilizers ammonium nitrate (NH_4NO_3) and ammonium sulfate (or ammoniun sulphate, $(NH_4)_2SO_4$) can be negligible below soil pH of about 7.2, but large losses from urea can occur in both acidic and basic soils. Within 24 hours after surface application, the extent of loss can account for 50 percent.

- A substantial loss of nitrogen from urea can be reduced or eliminated by soil incorporation. This can be done by tillage, such as plowing under or by disking, or by irrigation. Being highly soluble to water, the urea fertilizer will be carried into the soil and there behave just like other nitrogen fertilizers. Rainfall can substitute for irrigation.

- Apply singly or mix with the right fertilizers. The urea fertilizer can be applied alone or mixed with some other selected fertilizer materials. However, some blends should be immediately applied. Moreover, it should not be mixed with some fertilizers because a reaction will occur that will render one of the nutrients useless. Mixing of strongly basic materials with urea will result to loss of nitrogen as ammonia.

- Fertilizers that can be mixed with urea: calcium cyanamide, sulphate of potash, and sulphate of potash magnesia.

- Fertilizers that can be mixed with urea but not stored in excess of 2-3 days: Chilean nitrate, sulphate of ammonia, nitrogen magnesia, diammonium phosphate, basic slag, rock phosphate, and muriate of potash.

- Fertilizers that cannot be mixed with urea: Calcium nitrate, calcium ammonium nitrate or limestone ammonium nitrate, ammonium sulphate nitrate, nitropotash or potash ammonium nitrate, superphosphate, and triple superphosphate.

Coated Urea

Coated urea fertilizers are a group of controlled release fertilizers consisting of prills of urea coated in less-soluble chemicals such as sulfur, polymers, other products or a combination. These fertilizers mitigate some of the negative aspects of urea fertilization, such as fertilizer burn. The coatings release the urea either when penetrated by water, as with sulfur, or when broken down, as with polymers.

Urea is widely used as a nitrogen fertilizer. Its high solubility in water makes it useful for liquid application, and it has a much lower risk of causing fertilizer burn than other chemicals such as calcium cyanide or ammonium nitrate. However, the risk of fertilizer burn with urea can be unacceptably high in some situations, such as higher temperatures. The high water-solubility of urea can be disadvantageous in some cases as well.

One particular technique to mitigate these disadvantages has been to encapsulate prills of urea with less-soluble chemicals. These coatings permit the gradual release of urea in a controlled fashion, allowing for less-frequent applications.

Sulfur-coated Urea

Sulfur-coated urea, or SCU, fertilizers release nitrogen via water penetration through cracks and micropores in the coating. Once water penetrates through the coating, nitrogen release is rapid. The particles of fertilizer may in turn be sealed with wax to slow release further still, making microbial degradation necessary to permit water penetration. The size of fertilizer particles may also be varied in order to vary the time at which nitrogen release occurs. Sulfur-coated products typically range from 32% to 41% elemental nitrogen by weight. The sulfur coating process was originally developed by the Tennessee Valley Authority.

Sulfur-coated urea products can only be applied in granular form, and thus cannot be applied via liquid fertilization methods. It is not uncommon to find empty sulfur husks in turf once the nitrogen is released. Another disadvantage has to do with the relatively large particle size of sulfur-coated urea fertilizers, which makes their use on closely mown surfaces like putting greens impractical. However, more recently, materials with smaller particle sizes have been developed, permitting the use of sulfur-coated ureas on putting greens.

Polymer-coated Urea

Polymer-coated urea, also called plastic-coated urea, or PCU, fertilizers can permit a more precise rate of nitrogen release than sulfur-coated urea products. It's possible to produce polymer-coated products where the nitrogen release can be delayed for 10 months after application. The primary disadvantage of polymer-coated urea products is their relatively high cost compared to sulfur-coated urea.

Combination Products

Products that use a combination of sulfur-coating and polymer-coating also exist. Typically, these products consist of urea, coated with a layer of sulfur, which is in turn coated with a layer of polymer. Each coating layer is generally less than the normal thickness for the individual processes. These products are generally used as less-expensive alternatives to purely plastic-coated products, while still providing precise nitrogen release characteristics.

PREPARATION OF FERTILIZERS

Raw Materials

The fertilizers outlined here are compound fertilizers composed of primary fertilizers and secondary nutrients. These represent only one type of fertilizer, and other single nutrient types are also made. The raw materials, in solid form, can be supplied to fertilizer manufacturers in bulk quantities of thousands of tons, drum quantities, or in metal drums and bag containers.

Primary fertilizers include substances derived from nitrogen, phosphorus, and potassium. Various raw materials are used to produce these compounds. When ammonia is used as the nitrogen source in a fertilizer, one method of synthetic production requires the use of natural gas and air. The phosphorus component is made using sulfur, coal, and phosphate rock. The potassium source comes from potassium chloride, a primary component of potash.

Secondary nutrients are added to some fertilizers to help make them more effective. Calcium is obtained from limestone, which contains calcium carbonate, calcium sulphate, and calcium magnesium carbonate. The magnesium source in fertilizers is derived from dolomite. Sulfur is another material that is mined and added to fertilizers. Other mined materials include iron from ferrous sulfate, copper, and molybdenum from molybdenum oxide.

Manufacturing Process

Fully integrated factories have been designed to produce compound fertilizers. Depending on the actual composition of the end product, the production process will differ from manufacturer to manufacturer.

Nitrogen Fertilizer Component

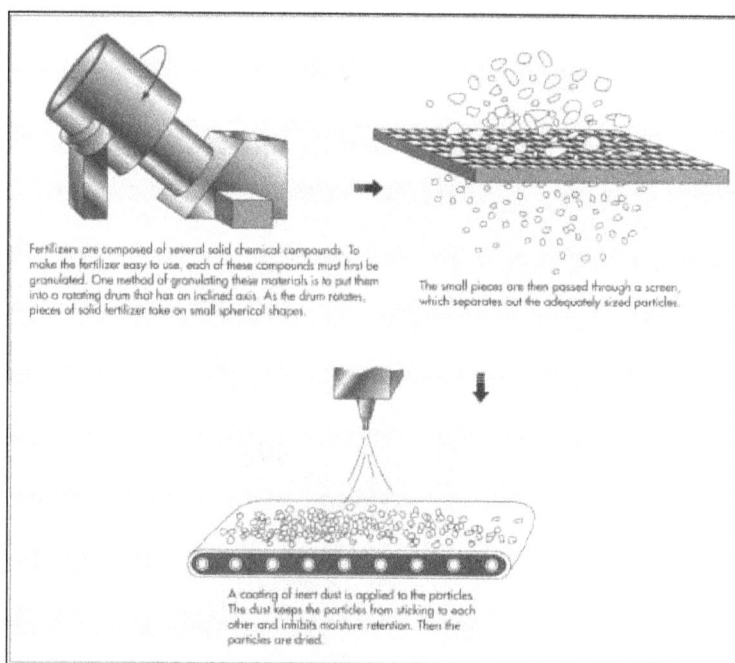

Fertilizers are composed of several solid chemical compounds. To make the fertilizer easy to use, each of these compounds must first be granulated. One method of granulating these materials is to put them into a rotating drum that has an inclined axis. As the drum rotates, pieces of solid fertilizer take on small spherical shapes.

The small pieces are then passed through a screen, which separates out the adequately sized particles.

A coating of inert dust is applied to the particles. The dust keeps the particles from sticking to each other and inhibits moisture retention. Then the particles are dried.

- Ammonia is one nitrogen fertilizer component that can be synthesized from in-expensive raw materials. Since nitrogen makes up a significant portion of the earth's atmosphere, a process was developed to produce ammonia from air. In this process, Fertilizernatural gas and steam are pumped into a large vessel. Next, air is pumped into the system, and oxygen is removed by the burning of natural gas and steam. This leaves primarily nitrogen, hydrogen, and carbon dioxide. The carbon dioxide is removed and ammonia is produced by introducing an electric current into the system. Catalysts such as magnetite (Fe_3O_4) have been used to improve the speed and efficiency of ammonia synthesis. Any impurities are removed from the ammonia, and it is stored in tanks until it is further processed.

- While ammonia itself is sometimes used as a fertilizer, it is often converted to other substances for ease of handling. Nitric acid is produced by first mixing ammonia and air in a tank. In the presence of a catalyst, a reaction occurs which converts the ammonia to nitric oxide. The nitric oxide is further reacted in the presence of water to produce nitric acid.

- Nitric acid and ammonia are used to make ammonium nitrate. This material is a good fertilizer component because it has a high concentration of nitrogen. The two materials are mixed together in a tank and a neutralization reaction occurs, producing ammonium nitrate. This material can then be stored until it is ready to be granulated and blended with the other fertilizer components.

Phosphorous Fertilizer Component

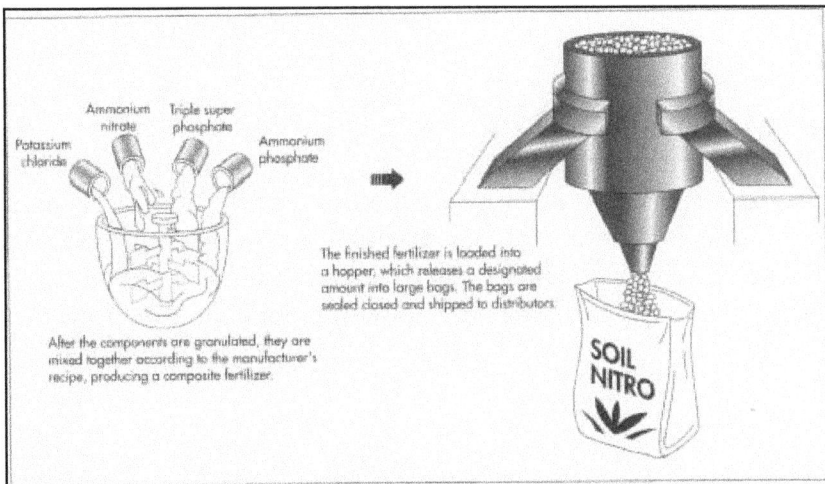

- To isolate phosphorus from phosphate rock, it is treated with sulfuric acid, producing phosphoric acid. Some of this material is reacted further with sulfuric acid and nitric acid to produce a triple superphosphate, an excellent source of phosphorous in solid form.

- Some of the phosphoric acid is also reacted with ammonia in a separate tank. This reaction results in ammonium phosphate, another good primary fertilizer.

Potassium Fertilizer Component

- Potassium chloride is typically supplied to fertilizer manufacturers in bulk. The manufacturer converts it into a more usable form by granulating it. This makes it easier to mix with other fertilizer components in the next step.

Granulating and Blending

- To produce fertilizer in the most usable form, each of the different compounds, ammonium nitrate, potassium chloride, ammonium phosphate, and triple superphosphate are granulated and blended together. One method of granulation involves putting the solid materials into a rotating drum which has an inclined axis. As the drum rotates, pieces of the solid fertilizer take on small spherical shapes. They are passed through a screen that separates out adequately sized particles. A coating of inert dust is then applied to the particles, keeping each one discrete and inhibiting moisture retention. Finally, the particles are dried, completing the granulation process.

- The different types of particles are blended together in appropriate proportions to produce a composite fertilizer. The blending is done in a large mixing drum that rotates a specific number of turns to produce the best mixture possible. After mixing, the fertilizer is emptied onto a conveyor belt, which transports it to the bagging machine.

Bagging

Fertilizers are typically supplied to farmers in large bags. To fill these bags the fertilizer is first delivered into a large hopper. An appropriate amount is released from the hopper into a bag that is held open by a clamping device. The bag is on a vibrating surface, which allows better packing. When filling is complete, the bag is transported upright to a machine that seals it closed. The bag is then conveyored to a palletizer, which stacks multiple bags, readying them for shipment to distributors and eventually to farmers.

Quality Control

To ensure the quality of the fertilizer that is produced, manufacturers monitor the product at each stage of production. The raw materials and the finished products are all subjected to a battery of physical and chemical tests to show that they meet the specifications previously developed. Some of the characteristics that are tested include pH, appearance, density, and melting point. Since fertilizer production is governmentally regulated, composition analysis tests are run on samples to determine total nitrogen content, phosphate content, and other elements affecting the chemical composition. Various other tests are also performed, depending on the specific nature of the fertilizer composition.

Byproducts/Waste

A relatively small amount of the nitrogen contained in fertilizers applied to the soil is actually assimilated into the plants. Much is washed into surrounding bodies of water or filters into the groundwater. This has added significant amounts of nitrates to the water that is consumed by the public. Some medical studies have suggested that certain disorders of the urinary and kidney systems are a result of excessive nitrates in drinking water. It is also thought that this is particularly harmful for babies and could even be potentially carcinogenic.

The nitrates that are contained in fertilizers are not thought to be harmful in themselves. However, certain bacteria in the soil convert nitrates into nitrite ions. Research has shown that when nitrite ions are ingested, they can get into the bloodstream. There, they bond with hemoglobin, a protein that is responsible for storing oxygen. When a nitrite ion binds with hemoglobin, it loses its ability to store oxygen, resulting in serious health problems.

ADVANTAGES AND DISADVANTAGES OF FERTILIZERS

Fertilizers are of mineral origin. They are nitrogenous, phosphatic, potassium salts and salts containing other elements. All the inorganic fertilizers have higher nutrient content compared to organic manures.

Advantages

- Rapid action;

- Constituent ratios defined.

Disadvantages

- Primarily made from non-renewable sources including fossil fuel;

- Provide nutrients to plants but nothing to sustain soil;

- Chance of over fertilization, that may damage crop and entire soil ecosystem;

- Leaching and move to rivers and sea causing eutrophication;

- Repeated application may lead to toxic build up of arsenic, cadmium etc in soil that may be present in fruits and vegetables;

- Long term use causes change in soil pH, may upset microbial ecosystem.

References

- Manure: britannica.com, Retrieved 29 March, 2019

- Piper, c.v.; pieters a.j. usda farmer's bulletin (ed.). Green manuring. Usda farmer's bulletin. Pp. 1250–1295. Retrieved feb 2, 2010

- Nitrogenous-fertilizers: agropedia.iitk.ac.in, Retrieved 14 July, 2019

- Barrett, j. (2008). Fcs soil science l3. Fet college series. Pearson education south africa. P. 70. Isbn 978-1-77025-114-4

- Compound-fertilizer: cropnutrition.com, Retrieved 30 April, 2019

- chicken manure adds to chesapeake bay pollution, group says". Wtop. December 28, 2011. Retrieved february 18, 2015

- Methods-of-composting: compostinstructions.com, Retrieved 29 June, 2019

- Morel, p.; guillemain, g. (2004). "assessment of the possible phytotoxicity of a substrate using an easy and representative biotest". Acta horticulturae (644): 417–423. Doi:10.17660/actahortic.2004.644.55

- Advantages-and-disadvantages-of: plantscience4u.com, Retrieved 11 January, 2019

5

Common Agrochemicals

The chemical compounds which are used in agriculture are termed as agrochemicals. Some of the commonly used agrochemicals are insecticides, pesticides, biopesticides, herbicides, bioherbicides and fungicides. The diverse aspects of these agrochemicals have been thoroughly discussed in this chapter.

Agrochemicals (agricultural chemicals, agrichemicals) are the various chemical products that are used in agriculture. In most cases, the term agrochemical refers to the broad range of pesticide chemicals, including insecticide chemicals, herbicide chemicals, fungicide chemicals, and nematicides chemicals (chemicals used to kill round worms). The term may also include synthetic fertilizers, hormones, and other chemical growth agents, as well as concentrated stores of raw animal manure.

Typically, agrochemicals are toxic and when stored in bulk storage systems may pose significant environmental risks, particularly in the event of accidental spills. As a result, in many countries, the use of agrochemicals has become highly regulated and government-issued permits for purchase and use of approved agrichemicals may be required. Significant penalties can result from misuse, including improper storage resulting in chemical leaks, chemical leaching, and chemical spills. Wherever these chemicals are used, proper storage facilities and labeling; emergency cleanup equipment; emergency cleanup procedures; safety equipment; as well as safety procedures for handling, application, and disposal are often subject to mandatory standards and regulations.

While agrochemicals increase plant and animal crop production, they can also damage the environment. Excessive use of fertilizers has led to the contamination of groundwater with nitrate, a chemical compound that in large concentrations is poisonous to humans and animals. In addition, the runoff (or leaching from the soil) of fertilizers into streams, lakes, and other surface waters (the aquasphere) can increase the growth of algae, which can have an adverse effect on the life-cycle of fish and other aquatic animals.

Pesticides that are sprayed on entire fields using equipment mounted on tractors, airplanes, or helicopters often drift away (due to wind or air convection patterns) from the targeted field, settling on nearby plants and animals. Some older pesticides, such as the powerful insecticide DDT (dichlorodiphenyltrichloroethane), remain active in the environment for many years, contaminating virtually all wildlife, well water, food, and even humans with whom it comes in contact. Although many of these pesticides have been banned, some newer pesticides still cause severe environmental damage.

Harmful Chemicals Identified by the United Nations Environment Program Governing Council:

- Aldrin: An insecticide used in soils to kill insects such as termites, grasshoppers, and western corn rootworm.

- Chlordane: An insecticide used to control termites and on a range of agricultural crops; a chemical that remains in the soil with a reported half-life of 1 year.

- Chlordecone: A synthetic chlorinated organic compound that is primarily used as an agricultural pesticide.

- Dichlorodiphenyltrichloroethane (DDT): Used as insecticide during WWII to protect against malaria and typhus; after the war, used as an agricultural insecticide; can persists in the soil for 10-15 years after application.

- Dieldrin: A pesticide used to control termites, textile pests, insect-borne diseases, and insects living in agricultural soils; half-life is approximately 5 years.

- Dioxins: By-products of high-temperature processes, such as incomplete combustion and pesticide production also emitted from the burning of hospital waste, municipal waste, and hazardous waste as well as automobile emissions, combustion of peat, coal, and wood.

- Endosulfans: Insecticides used to control pests on crops such coffee, cotton, rice, and sorghum and soybeans, tsetse flies, ectoparasites of cattle; also used as a wood preservative.

- Endrin: An insecticide sprayed on the leaves of crops, and used to control rodents; half-life is up to 12 years.

- Heptachlor: A pesticide primarily used to kill soil insects and termites, along with cotton insects, grasshoppers, other crop pests, and malaria-carrying mosquitoes.

- Hexabromocyclododecane (HBCD): A brominated flame retardant used as a thermal insulator in the building industry; persistent, toxic, and ecotoxic with bioaccumulative properties and long-range transport properties.

- Hexabromodiphenyl ether (hexaBDE) and heptabromodiphenyl ether: The main components of commercial octabromodiphenyl ether (octaBDE); highly persistent in the environment.

- Hexachlorobenzene: A fungicide used as a seed treatment, especially on wheat to control the fungal disease bunt; also a by-product produced during the manufacture of chlorinated solvents and other chlorinated compounds.

- α-Hexachlorocyclohexane (α-HCH) and β-hexachlorocyclohexane (β-HCH): Insecticides as well as by-products in the production of lindane; highly persistent in the water of colder regions.

- Lindane, also known as gamma-hexachlorocyclohexane, (γ-HCH), gammaxene, Gammallin, and sometimes incorrectly called benzene hexachloride (BHC): A chemical variant of hexachlorocyclohexane that has been used as an agricultural insecticide.

- Mirex: An insecticide used against ants and termites or as a flame retardant in plastics, rubber, and electrical goods; half-life is up to 10 years.

- Pentachlorobenzene (PeCB): A pesticide and also used in polychlorobiphenyl products, dyestuff carriers, as a fungicide, a flame retardant, and a chemical intermediate.

- Perfluorooctane sulfonic acid (PFOS) salts of the acid: Used in the production of fluoropolymers; extremely persistent in the environment through bioaccumulation and biomagnification.

- Polychlorinated biphenyls (PCBs): Used as heat exchange fluids in electrical transformers and capacitors; also used as additives in paint, carbonless copy paper, and plastics; a half-life up to 10 years.

- Polychlorinated dibenzofurans: By-products of high-temperature processes, such as incomplete combustion after waste incineration, pesticide production, and polychlorinated biphenyl production.

- Tetrabromodiphenyl ether (tetraBDE) and pentabromodiphenyl ether (pentaBDE): Industrial chemicals and the main components of commercial pentabromodiphenyl ether (pentaBDE).

- Toxaphene: An insecticide used on cotton, cereal, grain, fruits, nuts, and vegetables, as well as for tick and mite control in livestock; a half-life up to 12 years in soil.

There is now an awareness of the health hazards of pesticides and related chemicals due to the pioneering work that commenced in the latter half of the 20th century and has continued into the 21st century. These materials are carefully regulated, and the safety requirements for every pesticide product. Most fertilizers have been in an opposite category, considered useful, safe, and inert. These and other environmental effects have prompted the search for nonchemical methods of enhancing soil fertility and dealing with crop pests. These alternatives, however, are still emerging and are not yet in widespread use.

INSECTICIDES

Insecticides are substances used to kill insects. They include ovicides and larvicides used against insect eggs and larvae, respectively. Insecticides are used in agriculture, medicine, industry and by consumers. Insecticides are claimed to be a major factor behind the increase in the 20th-century's agricultural productivity. Nearly all insecticides have the potential to significantly alter ecosystems; many are toxic to humans and/or animals; some become concentrated as they spread along the food chain.

Insecticides can be classified into two major groups: systemic insecticides, which have residual or long term activity; and contact insecticides, which have no residual activity.

Furthermore, one can distinguish three types of insecticide: 1. Natural insecticides, such as nicotine, pyrethrum and neem extracts, made by plants as defenses against insects. 2. Inorganic insecticides, which are metals. 3. Organic insecticides, which are organic chemical compounds, mostly working by contact.

The mode of action describes how the pesticide kills or inactivates a pest. It provides another way of classifying insecticides. Mode of action is important in understanding whether an insecticide will be toxic to unrelated species, such as fish, birds and mammals.

Insecticides may be repellent or non-repellent. Social insects such as ants cannot detect non-repellents and readily crawl through them. As they return to the nest they take insecticide with them and transfer it to their nestmates. Over time, this eliminates all of the ants including the queen. This is slower than some other methods, but usually completely eradicates the ant colony.

Insecticides are distinct from non-insecticidal repellents, which repel but do not kill.

Type of Activity

Systemic insecticides become incorporated and distributed systemically throughout the whole plant. When insects feed on the plant, they ingest the insecticide. Systemic insecticides produced by transgenic plants are called plant-incorporated protectants (PIPs). For instance, a gene that codes for a specific Bacillus thuringiensis biocidal protein was introduced into corn (maize) and other species. The plant manufactures the protein, which kills the insect when consumed.

Contact insecticides are toxic to insects upon direct contact. These can be inorganic insecticides, which are metals and include the commonly used sulfur, and the less commonly used arsenates, copper and fluorine compounds. Contact insecticides can also be organic insecticides, i.e. organic chemical compounds, synthetically produced, and comprising the largest numbers of pesticides used today or they can be natural compounds like pyrethrum, neem oil etc. Contact insecticides usually have no residual activity.

Efficacy can be related to the quality of pesticide application, with small droplets, such as aerosols often improving performance.

Biological Pesticides

Many organic compounds are produced by plants for the purpose of defending the host plant from predation. A trivial case is tree rosin, which is a natural insecticide. Specifically, the production of oleoresin by conifer species is a component of the defense response against insect attack and fungal pathogen infection. Many fragrances, e.g. oil of wintergreen, are in fact antifeedants.

Four extracts of plants are in commercial use: pyrethrum, rotenone, neem oil, and various essential oils.

Other Biological Approaches

Plant-incorporated Protectants

Transgenic crops that act as insecticides began in 1996 with a genetically modified potato that produced the Cry protein, derived from the bacterium Bacillus thuringiensis, which is toxic to beetle larvae such as the Colorado potato beetle. The technique has been expanded to include the use of RNA interference RNAi that fatally silences crucial insect genes. RNAi likely evolved as a defense against viruses. Midgut cells in many larvae take up the molecules and help spread the signal. The technology can target only insects that have the silenced sequence, as was demonstrated when a particular RNAi affected only one of four fruit fly species. The technique is expected to replace many other insecticides, which are losing effectiveness due to the spread of pesticide resistance.

Enzymes

Many plants exude substances to repel insects. Premier examples are substances activated by the enzyme myrosinase. This enzyme converts glucosinolates to various compounds that are toxic to herbivorous insects. One product of this enzyme is allyl isothiocyanate, the pungent ingredient in horseradish sauces.

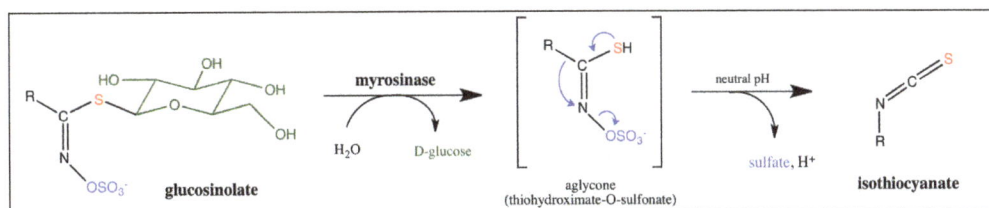

Biosynthesis of antifeedants by the action of myrosinase.

The myrosinase is released only upon crushing the flesh of horseradish. Since allyl isothiocyanate is harmful to the plant as well as the insect, it is stored in the harmless form of the glucosinolate, separate from the myrosinase enzyme.

Bacterial

Bacillus thuringiensis is a bacterial disease that affects Lepidopterans and some other insects. Toxins produced by strains of this bacterium are used as a larvicide against caterpillars, beetles, and mosquitoes. Toxins from *Saccharopolyspora spinosa* are isolated from fermentations and sold as Spinosad. Because these toxins have little effect on other organisms, they are considered more environmentally friendly than synthetic pesticides. The toxin from *B. thuringiensis* (Bt toxin) has been incorporated directly into plants through the use of genetic engineering. Other biological insecticides include products based on entomopathogenic fungi (e.g., *Beauveria bassiana*, *Metarhizium anisopliae*), nematodes (e.g., *Steinernema feltiae*) and viruses (e.g., *Cydia pomonella* granulovirus).

Synthetic Insecticide and Natural Insecticides

A major emphasis of organic chemistry is the development of chemical tools to enhance agricultural productivity. Insecticides represent a major area of emphasis. Many of the major insecticides are inspired by biological analogues. Many others are completely alien to nature.

Organochlorides

The best known organochloride, DDT, was created by Swiss scientist Paul Müller. For this discovery, he was awarded the 1948 Nobel Prize for Physiology or Medicine. DDT was introduced in 1944. It functions by opening sodium channels in the insect's nerve cells. The contemporaneous rise of the chemical industry facilitated large-scale production of DDT and related chlorinated hydrocarbons.

Organophosphates and Carbamates

Organophosphates are another large class of contact insecticides. These also target the insect's nervous system. Organophosphates interfere with the enzymes acetylcholinesterase and other cholinesterases, disrupting nerve impulses and killing or disabling the insect. Organophosphate insecticides and chemical warfare nerve agents (such as sarin, tabun, soman, and VX) work in the same way. Organophosphates have a cumulative toxic effect to wildlife, so multiple exposures to the chemicals amplifies the toxicity. In the US, organophosphate use declined with the rise of substitutes.

Carbamate insecticides have similar mechanisms to organophosphates, but have a much shorter duration of action and are somewhat less toxic.

Pyrethroids

Pyrethroid pesticides mimic the insecticidal activity of the natural compound pyrethrum, the biopesticide found in pyrethrins. These compounds are nonpersistent sodium channel modulators and are less toxic than organophosphates and carbamates. Compounds in this group are often applied against household pests.

Neonicotinoids

Neonicotinoids are synthetic analogues of the natural insecticide nicotine (with much lower acute mammalian toxicity and greater field persistence). These chemicals are acetylcholine receptor agonists. They are broad-spectrum systemic insecticides, with rapid action (minutes-hours). They are applied as sprays, drenches, seed and soil treatments. Treated insects exhibit leg tremors, rapid wing motion, stylet withdrawal (aphids), disoriented movement, paralysis and death. Imidacloprid may be the most common. It has recently come under scrutiny for allegedly pernicious effects on honeybees and its potential to increase the susceptibility of rice to planthopper attacks.

Ryanoids

Ryanoids are synthetic analogues with the same mode of action as ryanodine, a naturally occurring insecticide extracted from *Ryania speciosa* (Flacourtiaceae). They bind to calcium channels in cardiac and skeletal muscle, blocking nerve transmission. The first insecticide from this class to be registered was Rynaxypyr, generic name chlorantraniliprole.

Insect Growth Regulators

Insect growth regulator (IGR) is a term coined to include insect hormone mimics and an earlier class of chemicals, the benzoylphenyl ureas, which inhibit chitin (exoskeleton) biosynthesis in insects Diflubenzuron is a member of the latter class, used primarily to control caterpillars that are pests. The most successful insecticides in this class are the juvenoids (juvenile hormone analogues). Of these, methoprene is most widely used. It has no observable acute toxicity in rats and is approved by World Health Organization (WHO) for use in drinking water cisterns to combat malaria. Most of its uses are to combat insects where the adult is the pest, including mosquitoes, several fly species, and fleas. Two very similar products, hydroprene and kinoprene, are used for controlling species such as cockroaches and white flies. Methoprene was registered with the EPA in 1975. Virtually no reports of resistance have been filed. A more recent type of IGR is the ecdysone agonist tebufenozide (MIMIC), which is used in forestry and other applications for control of caterpillars, which are far more sensitive to its hormonal effects than other insect orders.

Environmental Harm

Effects on Nontarget Species

Some insecticides kill or harm other creatures in addition to those they are intended to kill. For example, birds may be poisoned when they eat food that was recently sprayed with insecticides or when they mistake an insecticide granule on the ground for food and eat it. Sprayed insecticide may drift from the area to which it is applied and into wildlife areas, especially when it is sprayed aerially.

DDT

The development of DDT was motivated by desire to replace more dangerous or less effective alternatives. DDT was introduced to replace lead and arsenic-based compounds, which were in widespread use in the early 1940s.

DDT was brought to public attention by Rachel Carson's book *Silent Spring*. One side-effect of DDT is to reduce the thickness of shells on the eggs of predatory birds. The shells sometimes become too thin to be viable, reducing bird populations. This occurs with DDT and related compounds due to the process of bioaccumulation, wherein the chemical, due to its stability and fat solubility, accumulates in organisms' fatty tissues. Also, DDT may biomagnify, which causes progressively higher concentrations in the body fat of animals farther up the food chain. The near-worldwide ban on agricultural use of DDT and related chemicals has allowed some of these birds, such as the peregrine falcon, to recover in recent years. A number of organochlorine pesticides have been banned from most uses worldwide. Globally they are controlled via the Stockholm Convention on persistent organic pollutants. These include: aldrin, chlordane, DDT, dieldrin, endrin, heptachlor, mirex and toxaphene.

Pollinator Decline

Insecticides can kill bees and may be a cause of pollinator decline, the loss of bees that pollinate plants, and colony collapse disorder (CCD), in which worker bees from a beehive or Western honey bee colony abruptly disappear. Loss of pollinators means a reduction in crop yields. Sublethal doses of insecticides (i.e. imidacloprid and other neonicotinoids) affect bee foraging behavior. However, research into the causes of CCD was inconclusive as of June 2007.

Bird Decline

Besides the effects of direct consumption of insecticides, populations of insectivorous birds decline due to the collapse of their prey populations. Spraying of especially wheat and corn in Europe is believed to have caused an 80 per cent decline in flying insects, which in turn has reduced local bird populations by a third to two thirds.

Alternatives

Instead of using chemical insecticides to avoid crop damage caused by insects, there are many alternative options available now that can protect farmers from major economic losses. Some of them are:

1. Breeding crops resistant, or at least less susceptible, to pest attacks.

2. Releasing predators, parasitoids, or pathogens to control pest populations as a form of biological control.

3. Chemical control like releasing pheromones into the field to confuse the insects into not being able to find mates and reproduce.

4. Integrated Pest Management- using multiple techniques in tandem to achieve optimal results.

5. Push-pull technique- intercropping with a "push" crop that repels the pest, and planting a "pull" crop on the boundary that attracts and traps it.

PESTICIDES

Pesticides are substances that are meant to control pests, including weeds. The term pesticide includes all of the following: herbicide, insecticides (which may include insect growth regulators, termiticides, etc.) nematicide, molluscicide, piscicide, avicide, rodenticide, bactericide, insect repellent, animal repellent, antimicrobial, and fungicide. The most common of these are herbicides which account for approximately 80% of all pesticide use. Most pesticides are intended to serve as plant protection products (also known as crop protection products), which in general, protect plants from weeds, fungi, or insects.

In general, a pesticide is a chemical or biological agent (such as a virus, bacterium, or fungus) that deters, incapacitates, kills, or otherwise discourages pests. Target pests can include insects, plant pathogens, weeds, molluscs, birds, mammals, fish, nematodes (roundworms), and microbes that destroy property, cause nuisance, or spread disease, or are disease vectors. Along with these benefits, pesticides also have drawbacks, such as potential toxicity to humans and other species.

Type of pesticide	Target pest group
Algicides or algaecides	Algae
Avicides	Birds
Bactericides	Bacteria
Fungicides	Fungi and oomycetes
Herbicides	Plant
Insecticides	Insects
Miticides or acaricides	Mites
Molluscicides	Snails

Nematicides	Nematodes
Rodenticides	Rodents
Slimicides	Algae, Bacteria, Fungi, and Slime molds
Virucides	Viruses

The Food and Agriculture Organization (FAO) has defined *pesticide* as:

Any substance or mixture of substances intended for preventing, destroying, or controlling any pest, including vectors of human or animal disease, unwanted species of plants or animals, causing harm during or otherwise interfering with the production, processing, storage, transport, or marketing of food, agricultural commodities, wood and wood products or animal feedstuffs, or substances that may be administered to animals for the control of insects, arachnids, or other pests in or on their bodies. The term includes substances intended for use as a plant growth regulator, defoliant, desiccant, or agent for thinning fruit or preventing the premature fall of fruit. Also used as substances applied to crops either before or after harvest to protect the commodity from deterioration during storage and transport.

Pesticides can be classified by target organism (e.g., herbicides, insecticides, fungicides, rodenticides, and pediculicides), chemical structure (e.g., organic, inorganic, synthetic, or biological (biopesticide), although the distinction can sometimes blur), and physical state (e.g. gaseous (fumigant)). Biopesticides include microbial pesticides and biochemical pesticides. Plant-derived pesticides, or "botanicals", have been developing quickly. These include the pyrethroids, rotenoids, nicotinoids, and a fourth group that includes strychnine and scilliroside.

Many pesticides can be grouped into chemical families. Prominent insecticide families include organochlorines, organophosphates, and carbamates. Organochlorine hydrocarbons (e.g., DDT) could be separated into dichlorodiphenylethanes, cyclodiene compounds, and other related compounds. They operate by disrupting the sodium/potassium balance of the nerve fiber, forcing the nerve to transmit continuously. Their toxicities vary greatly, but they have been phased out because of their persistence and potential to bioaccumulate. Organophosphate and carbamates largely replaced organochlorines. Both operate through inhibiting the enzyme acetylcholinesterase, allowing acetylcholine to transfer nerve impulses indefinitely and causing a variety of symptoms such as weakness or paralysis. Organophosphates are quite toxic to vertebrates and have in some cases been replaced by less toxic carbamates. Thiocarbamate and dithiocarbamates are subclasses of carbamates. Prominent families of herbicides include phenoxy and benzoic acid herbicides (e.g. 2,4-D), triazines (e.g., atrazine), ureas (e.g., diuron), and Chloroacetanilides (e.g., alachlor). Phenoxy compounds tend to selectively kill broad-leaf weeds rather than grasses. The phenoxy and benzoic acid herbicides function similar to plant growth hormones, and grow cells without normal cell division, crushing the plant's nutrient transport system. Triazines interfere with photosynthesis. Many commonly used pesticides are not included in these families, including glyphosate.

The application of pest control agents is usually carried out by dispersing the chemical in a (often hydrocarbon-based) solvent-surfactant system to give a homogeneous preparation. A virus lethality study performed in 1977 demonstrated that a particular pesticide did not increase the lethality of the virus, however combinations which included some surfactants and the solvent clearly showed that pretreatment with them markedly increased the viral lethality in the test mice.

Pesticides can be classified based upon their biological mechanism function or application method. Most pesticides work by poisoning pests. A systemic pesticide moves inside a plant following absorption by the plant. With insecticides and most fungicides, this movement is usually upward (through the xylem) and outward. Increased efficiency may be a result. Systemic insecticides, which poison pollen and nectar in the flowers, may kill bees and other needed pollinators.

In 2010, the development of a new class of fungicides called paldoxins was announced. These work by taking advantage of natural defense chemicals released by plants called phytoalexins, which fungi then detoxify using enzymes. The paldoxins inhibit the fungi's detoxification enzymes. They are believed to be safer and greener.

Uses

Pesticides are used to control organisms that are considered to be harmful. For example, they are used to kill mosquitoes that can transmit potentially deadly diseases like West Nile virus, yellow fever, and malaria. They can also kill bees, wasps or ants that can cause allergic reactions. Insecticides can protect animals from illnesses that can be caused by parasites such as fleas. Pesticides can prevent sickness in humans that could be caused by moldy food or diseased produce. Herbicides can be used to clear roadside weeds, trees, and brush. They can also kill invasive weeds that may cause environmental damage. Herbicides are commonly applied in ponds and lakes to control algae and plants such as water grasses that can interfere with activities like swimming and fishing and cause the water to look or smell unpleasant. Uncontrolled pests such as termites and mold can damage structures such as houses. Pesticides are used in grocery stores and food storage facilities to manage rodents and insects that infest food such as grain. Each use of a pesticide carries some associated risk. Proper pesticide use decreases these associated risks to a level deemed acceptable by pesticide regulatory agencies such as the United States Environmental Protection Agency (EPA) and the Pest Management Regulatory Agency (PMRA) of Canada.

DDT, sprayed on the walls of houses, is an organochlorine that has been used to fight malaria since the 1950s. Recent policy statements by the World Health Organization have given stronger support to this approach. However, DDT and other organochlorine pesticides have been banned in most countries worldwide because of their persistence in the environment and human toxicity. DDT use is not always effective, as resistance to DDT was identified in Africa as early as 1955, and by 1972 nineteen species of mosquito worldwide were resistant to DDT.

Amount Used

In 2006 and 2007, the world used approximately 2.4 megatonnes (5.3×10^9 lb) of pesticides, with herbicides constituting the biggest part of the world pesticide use at 40%, followed by insecticides (17%) and fungicides (10%). In 2006 and 2007 the U.S. used approximately 0.5 megatonnes (1.1×10^9 lb) of pesticides, accounting for 22% of the world total, including 857 million pounds (389 kt) of conventional pesticides, which are used in the agricultural sector (80% of conventional pesticide use) as well as the industrial, commercial, governmental and home and garden sectors. The state of California alone used 117 million pounds. Pesticides are also found in majority of U.S. households with 88 million out of the 121.1 million households indicating that they use some form of pesticide in 2012. As of 2007, there were more than 1,055 active ingredients registered as pesticides, which yield over 20,000 pesticide products that are marketed in the United States.

The US used some 1 kg (2.2 pounds) per hectare of arable land compared with: 4.7 kg in China, 1.3 kg in the UK, 0.1 kg in Cameroon, 5.9 kg in Japan and 2.5 kg in Italy. Insecticide use in the US has declined by more than half since 1980 (.6%/yr), mostly due to the near phase-out of organophosphates. In corn fields, the decline was even steeper, due to the switchover to transgenic Bt corn.

For the global market of crop protection products, market analysts forecast revenues of over 52 billion US$ in 2019.

Benefits

Pesticides can save farmers' money by preventing crop losses to insects and other pests; in the U.S., farmers get an estimated fourfold return on money they spend on pesticides. One study found that not using pesticides reduced crop yields by about 10%. Another study, conducted in 1999, found that a ban on pesticides in the United States may result in a rise of food prices, loss of jobs, and an increase in world hunger.

There are two levels of benefits for pesticide use, primary and secondary. Primary benefits are direct gains from the use of pesticides and secondary benefits are effects that are more long-term.

Primary Benefits

Controlling pests and plant disease vectors:

- Improved crop yields,

- Improved crop/livestock quality,

- Invasive species controlled,

Controlling human/livestock disease vectors and nuisance organisms:

- Human lives saved and disease reduced. Diseases controlled include malaria, with millions of lives having been saved or enhanced with the use of DDT alone.

- Animal lives saved and disease reduced.

Controlling organisms that harm other human activities and structures:

- Drivers view unobstructed.

- Tree/brush/leaf hazards prevented.

- Wooden structures protected.

Monetary

In one study, it was estimated that for every dollar ($1) that is spent on pesticides for crops can yield up to four dollars ($4) in crops saved. This means based that, on the amount of money spent per year on pesticides, $10 billion, there is an additional $40 billion savings in crop that would be lost due to damage by insects and weeds. In general, farmers benefit from having an increase in crop yield and from being able to grow a variety of crops throughout the year. Consumers of agricultural products also benefit from being able to afford the vast quantities of produce available year-round.

Costs

On the cost side of pesticide use there can be costs to the environment, costs to human health, as well as costs of the development and research of new pesticides.

Health Effects

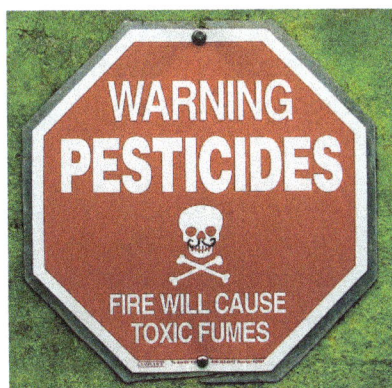

A sign warning about potential pesticide exposure.

Pesticides may cause acute and delayed health effects in people who are exposed. Pesticide exposure can cause a variety of adverse health effects, ranging from simple irritation of the skin and eyes to more severe effects such as affecting the nervous system, mimicking hormones causing reproductive problems, and also causing cancer. A 2007 systematic review found that "most studies on non-Hodgkin lymphoma and leukemia showed positive associations with pesticide exposure" and thus concluded that cosmetic use of pesticides should be decreased. There is substantial evidence of associations between organophosphate insecticide exposures and neurobehavioral alterations. Limited evidence also exists for other negative outcomes from pesticide exposure including neurological, birth defects, and fetal death.

The American Academy of Pediatrics recommends limiting exposure of children to pesticides and using safer alternatives.

Owing to inadequate regulation and safety precautions, 99% of pesticide related deaths occur in developing countries that account for only 25% of pesticide usage.

One study found pesticide self-poisoning the method of choice in one third of suicides worldwide, and recommended, among other things, more restrictions on the types of pesticides that are most harmful to humans.

A 2014 epidemiological review found associations between autism and exposure to certain pesticides, but noted that the available evidence was insufficient to conclude that the relationship was causal.

Large quantities of presumably nontoxic petroleum oil by-products are introduced into the environment as pesticide dispersal agents and emulsifiers. A 1976 study found that an increase in viral lethality with a concomitant influence on the liver and central nervous system occurs in young mice previously primed with such chemicals.

The World Health Organization and the UN Environment Programme estimate that each year, 3 million workers in agriculture in the developing world experience severe poisoning from pesticides, about 18,000 of whom die. According to one study, as many as 25 million workers in developing countries may suffer mild pesticide poisoning yearly. There are several careers aside from agriculture that may also put individuals at risk of health effects from pesticide exposure including pet groomers, groundskeepers, and fumigators.

Pesticide use is widespread in Latin America, as around US$3 billion are spent each year in the region. It has been recorded that pesticide poisonings have been increasing each year for the past two decades. It was estimated that 50–80% of the cases are unreported. It is indicated by studies that organophosphate and carbamate insecticides are the most frequent source of pesticide poisoning.

Environmental Effects

Pesticide use raises a number of environmental concerns. Over 98% of sprayed insecticides and 95% of herbicides reach a destination other than their target species, including non-target species, air, water and soil. Pesticide drift occurs when pesticides suspended in the air as particles are carried by wind to other areas, potentially contaminating them. Pesticides are one of the causes of water pollution, and some pesticides are persistent organic pollutants and contribute to soil and flower (pollen, nectar) contamination.

In addition, pesticide use reduces biodiversity, contributes to pollinator decline, destroys habitat (especially for birds), and threatens endangered species. Pests can develop a resistance to the pesticide (pesticide resistance), necessitating a new pesticide. Alternatively a greater dose of the pesticide can be used to counteract the resistance, although this will cause a worsening of the ambient pollution problem.

The Stockholm Convention on Persistent Organic Pollutants, listed 9 of the 12 most dangerous and persistent organic chemicals that were (now mostly obsolete) organochlorine pesticides. Since chlorinated hydrocarbon pesticides dissolve in fats and are not excreted, organisms tend to retain them almost indefinitely. Biological magnification is the process whereby these chlorinated hydrocarbons (pesticides) are more concentrated at each level of the food chain. Among marine animals, pesticide concentrations are higher in carnivorous fishes, and even more so in the fish-eating birds and mammals at the top of the ecological pyramid. Global distillation is the process whereby pesticides are transported from warmer to colder regions of the Earth, in particular the Poles and mountain tops. Pesticides that evaporate into the atmosphere at relatively high temperature can be carried considerable distances (thousands of kilometers) by the wind to an area of lower temperature, where they condense and are carried back to the ground in rain or snow.

In order to reduce negative impacts, it is desirable that pesticides be degradable or at least quickly deactivated in the environment. Such loss of activity or toxicity of pesticides is due to both innate chemical properties of the compounds and environmental processes or conditions. For example, the presence of halogens within a chemical structure often slows down degradation in an aerobic environment. Adsorption to soil may retard pesticide movement, but also may reduce bioavailability to microbial degraders.

Economics

Harm	Annual US cost
Public health	$1.1 billion
Pesticide resistance in pest	$1.5 billion
Crop losses caused by pesticides	$1.4 billion
Bird losses due to pesticides	$2.2 billion
Groundwater contamination	$2.0 billion
Other costs	$1.4 billion
Total costs	$9.6 billion

In one study, the human health and environmental costs due to pesticides in the United States was estimated to be $9.6 billion: offset by about $40 billion in increased agricultural production.

Additional costs include the registration process and the cost of purchasing pesticides: which are typically borne by agrichemical companies and farmers respectively. The registration process can take several years to complete (there are 70 different types of field test) and can cost $50–70 million for a single pesticide. At the beginning of the 21st century, the United States spent approximately $10 billion on pesticides annually.

Alternatives

Alternatives to pesticides are available and include methods of cultivation, use of biological pest controls (such as pheromones and microbial pesticides), genetic engineering, and methods of interfering with insect breeding. Application of composted yard waste has also been used as a way of controlling pests. These methods are becoming increasingly popular and often are safer than traditional chemical pesticides. In addition, EPA is registering reduced-risk conventional pesticides in increasing numbers.

Cultivation practices include polyculture (growing multiple types of plants), crop rotation, planting crops in areas where the pests that damage them do not live, timing planting according to when pests will be least problematic, and use of trap crops that attract pests away from the real crop. Trap crops have successfully controlled pests in some commercial agricultural systems while reducing pesticide usage; however, in many other systems, trap crops can fail to reduce pest densities at a commercial scale, even when the trap crop works in controlled experiments. In the U.S., farmers have had success controlling insects by spraying with hot water at a cost that is about the same as pesticide spraying.

Release of other organisms that fight the pest is another example of an alternative to pesticide use. These organisms can include natural predators or parasites of the pests. Biological pesticides based on entomopathogenic fungi, bacteria and viruses cause disease in the pest species can also be used.

Interfering with insects' reproduction can be accomplished by sterilizing males of the target species and releasing them, so that they mate with females but do not produce offspring. This technique was first used on the screwworm fly in 1958 and has since been used with the medfly, the tsetse fly, and the gypsy moth. However, this can be a costly, time consuming approach that only works on some types of insects.

Push-Pull Strategy

The term "push-pull" was established in 1987 as an approach for integrated pest management (IPM). This strategy uses a mixture of behavior-modifying stimuli to manipulate the distribution and abundance of insects. "Push" means the insects are repelled or deterred away from whatever resource that is being protected. "Pull" means that certain stimuli (semiochemical stimuli, pheromones, food additives, visual stimuli, genetically altered plants, etc.) are used to attract pests to trap crops where they will be killed. There are numerous different components involved in order to implement a Push-Pull Strategy in IPM.

Many case studies testing the effectiveness of the push-pull approach have been done across the world. The most successful push-pull strategy was developed in Africa for subsistence farming. Another successful case study was performed on the control of *Helicoverpa* in cotton crops in Australia. In Europe, the Middle East, and the United States, push-pull strategies were successfully used in the controlling of *Sitona lineatus* in bean fields.

Some advantages of using the push-pull method are less use of chemical or biological materials and better protection against insect habituation to this control method. Some disadvantages of the push-pull strategy is that if there is a lack of appropriate knowledge of behavioral and chemical ecology of the host-pest interactions then this method becomes unreliable. Furthermore, because the push-pull method is not a very popular method of IPM operational and registration costs are higher.

Effectiveness

Some evidence shows that alternatives to pesticides can be equally effective as the use of chemicals. For example, Sweden has halved its use of pesticides with hardly any reduction in crops. In Indonesia, farmers have reduced pesticide use on rice fields by 65% and experienced a 15% crop increase. A study of Maize fields in northern Florida found that the application of composted yard waste with high carbon to nitrogen ratio to agricultural fields was highly effective at reducing the population of plant-parasitic nematodes and increasing crop yield, with yield increases ranging from 10% to 212%; the observed effects were long-term, often not appearing until the third season of the study.

However, pesticide resistance is increasing. In the 1940s, U.S. farmers lost only 7% of their crops to pests. Since the 1980s, loss has increased to 13%, even though more pesticides are being used. Between 500 and 1,000 insect and weed species have developed pesticide resistance since 1945.

Types

Pesticides are often referred to according to the type of pest they control. Pesticides can also be considered as either biodegradable pesticides, which will be broken down by microbes and other living beings into harmless compounds, or persistent pesticides, which may take months or years before they are broken down: it was the persistence of DDT, for example, which led to its accumulation in the food chain and its killing of birds of prey at the top of the food chain. Another way to think about pesticides is to consider those that are chemical pesticides are derived from a common source or production method.

Biopesticides

Biopesticides are certain types of pesticides derived from such natural materials as animals, plants, bacteria, and certain minerals. For example, canola oil and baking soda have pesticidal applications and are considered biopesticides. Biopesticides fall into three major classes:

- Microbial pesticides which consist of bacteria, entomopathogenic fungi or viruses (and sometimes includes the metabolites that bacteria or fungi produce). Entomopathogenic nematodes are also often classed as microbial pesticides, even though they are multi-cellular.

- Biochemical pesticides or herbal pesticides are naturally occurring substances that control (or monitor in the case of pheromones) pests and microbial diseases.

- Plant-incorporated protectants (PIPs) have genetic material from other species incorporated into their genetic material (*i.e.* GM crops). Their use is controversial, especially in many European countries.

Classified by Type of Pest

Pesticides that are related to the type of pests are:

Type	Action
Algicides	Control algae in lakes, canals, swimming pools, water tanks, and other sites
Antifouling agents	Kill or repel organisms that attach to underwater surfaces, such as boat bottoms
Antimicrobials	Kill microorganisms (such as bacteria and viruses)
Attractants	Attract pests (for example, to lure an insect or rodent to a trap). (However, food is not considered a pesticide when used as an attractant.)
Biopesticides	Biopesticides are certain types of pesticides derived from such natural materials as animals, plants, bacteria, and certain minerals
Biocides	Kill microorganisms
Disinfectants and sanitizers	Kill or inactivate disease-producing microorganisms on inanimate objects
Fungicides	Kill fungi (including blights, mildews, molds, and rusts)
Fumigants	Produce gas or vapor intended to destroy pests in buildings or soil
Herbicides	Kill weeds and other plants that grow where they are not wanted
Insecticides	Kill insects and other arthropods
Miticides	Kill mites that feed on plants and animals

Microbial pesticides	Microorganisms that kill, inhibit, or out compete pests, including insects or other microorganisms
Molluscicides	Kill snails and slugs
Nematicides	Kill nematodes (microscopic, worm-like organisms that feed on plant roots)
Ovicides	Kill eggs of insects and mites
Pheromones	Biochemicals used to disrupt the mating behavior of insects
Repellents	Repel pests, including insects (such as mosquitoes) and birds
Rodenticides	Control mice and other rodents
Slimicides	Kill slime-producing microorganisms such as algae, bacteria, fungi, and slime molds

Further Types

The term pesticide also include these substances:

1. Defoliants: Cause leaves or other foliage to drop from a plant, usually to facilitate harvest. 2. Desiccants: Promote drying of living tissues, such as unwanted plant tops. 3. Insect growth regulators: Disrupt the molting, maturity from pupal stage to adult, or other life processes of insects. 4. Plant growth regulators: Substances (excluding fertilizers or other plant nutrients) that alter the expected growth, flowering, or reproduction rate of plants. 5. Wood preservatives: They are used to make wood resistant to insects, fungus, and other pests.

Residue

Pesticide residue refers to the pesticides that may remain on or in food after they are applied to food crops. The maximum allowable levels of these residues in foods are often stipulated by regulatory bodies in many countries. Regulations such as pre-harvest intervals also often prevent harvest of crop or livestock products if recently treated in order to allow residue concentrations to decrease over time to safe levels before harvest. Exposure of the general population to these residues most commonly occurs through consumption of treated food sources, or being in close contact to areas treated with pesticides such as farms or lawns.

Many of these chemical residues, especially derivatives of chlorinated pesticides, exhibit bioaccumulation which could build up to harmful levels in the body as well as in the environment. Persistent chemicals can be magnified through the food chain and have been detected in products ranging from meat, poultry, and fish, to vegetable oils, nuts, and various fruits and vegetables.

Pesticide contamination in the environment can be monitored through bioindicators such as bee pollinators.

BIOPESTICIDE

Biopesticides, a contraction of 'biological pesticides', include several types of pest management intervention: through predatory, parasitic, or chemical relationships. The term has been associated historically with biological control – and by implication – the manipulation of living organisms. Regulatory positions can be influenced by public perceptions, thus:

- In the EU, biopesticides have been defined as "a form of pesticide based on micro-organisms or natural products".

- The US EPA states that they "include naturally occurring substances that control pests (biochemical pesticides), microorganisms that control pests (microbial pesticides), and pesticidal substances produced by plants containing added genetic material (plant-incorporated protectants) or PIPs".

They are obtained from organisms including plants, bacteria and other microbes, fungi, nematodes, *etc.* They are often important components of integrated pest management (IPM) programmes, and have received much practical attention as substitutes to synthetic chemical plant protection products (PPPs).

Types

Biopesticides can be classified into these classes:

- Microbial pesticides which consist of bacteria, entomopathogenic fungi or viruses (and sometimes includes the metabolites that bacteria or fungi produce). Entomopathogenic nematodes are also often classed as microbial pesticides, even though they are multi-cellular.

- Bio-derived chemicals. Four groups are in commercial use: pyrethrum, rotenone, neem oil, and various essential oils are naturally occurring substances that control (or monitor in the case of pheromones) pests and microbial diseases.

- Plant-incorporated protectants (PIPs) have genetic material from other species incorporated into their genetic material (*i.e.* GM crops). Their use is controversial, especially in many European countries.

- RNAi pesticides, some of which are topical and some of which are absorbed by the crop.

Biopesticides have usually no known function in photosynthesis, growth or other basic aspects of plant physiology. Instead, they are active against biological pests. Many chemical compounds have been identified that are produced by plants to protect them from pests so they are called antifeedants. These materials are biodegradable and renewable alternatives, which can be economical for practical use. Organic farming systems embraces this approach to pest control.

RNA

RNA interference is under study for possible use as a spray-on insecticide by multiple companies, including Monsanto, Syngenta, and Bayer. Such sprays do not modify the genome of the target plant. The RNA could be modified to maintain its effectiveness as target species evolve tolerance to the original. RNA is a relatively fragile molecule that generally degrades within days or weeks of application. Monsanto estimated costs to be on the order of $5/acre.

RNAi has been used to target weeds that tolerate Monsanto's Roundup herbicide. RNAi mixed with a silicone surfactant that let the RNA molecules enter air-exchange holes in the plant's surface that disrupted the gene for tolerance, affecting it long enough to let the herbicide work. This strategy would allow the continued use of glyphosate-based herbicides, but would not per se assist a herbicide rotation strategy that relied on alternating Roundup with others.

They can be made with enough precision to kill some insect species, while not harming others. Monsanto is also developing an RNA spray to kill potato beetles. One challenge is to make it linger on the plant for a week, even if it's raining. The Potato beetle has become resistant to more than 60 conventional insecticides.

Monsanto lobbied the U.S. EPA to exempt RNAi pesticide products from any specific regulations (beyond those that apply to all pesticides) and be exempted from rodent toxicity, allergenicity and residual environmental testing. In 2014 an EPA advisory group found little evidence of a risk to people from eating RNA.

However, in 2012, the Australian Safe Food Foundation posited that the RNA trigger designed to change the starch content of wheat might interfere with the gene for a human liver enzyme. Supporters countered that RNA does not appear to make it past human saliva or stomach acids. The US National Honey Bee Advisory Board told EPA that using RNAi would put natural systems at "the epitome of risk". The beekeepers cautioned that pollinators could be hurt by unintended effects and that the genomes of many insects are still unknown. Other unassessed risks include ecological (given the need for sustained presence for herbicide and other applications) and the possible for RNA drift across species boundaries.

Monsanto has invested in multiple companies for their RNA expertise, including Beeologics (for RNA that kills a parasitic mite that infests hives and for manufacturing technology) and Preceres (nanoparticle lipidoid coatings) and licensed technology from Alnylam and Tekmira. In 2012 Syngenta acquired Devgen, a European RNA partner. Startup Forrest Innovations is investigating RNAi as a solution to citrus greening disease that in 2014 caused 22 percent of oranges in Florida to fall off the trees.

Examples of RNA

Bacillus thuringiensis, a bacterial disease of Lepidoptera, Coleoptera and Diptera, is a well-known insecticide example. The toxin from *B. thuringiensis* (Bt toxin) has been incorporated directly into plants through the use of genetic engineering. The use of Bt Toxin is particularly controversial. Its manufacturers claim it has little effect on other organisms, and is more environmentally friendly than synthetic pesticides.

Other microbial control agents include products based on:

- Entomopathogenic fungi (*e.g. Beauveria bassiana, Isaria fumosorosea, Lecanicillium* and *Metarhizium* spp.).

- Plant disease control agents: include *Trichoderma* spp. and *Ampelomyces quisqualis* (a hyper-parasite of grape powdery mildew); *Bacillus subtilis* is also used to control plant pathogens.

- Beneficial nematodes attacking insect (*e.g. Steinernema feltiae*) or slug (*e.g. Phasmarhabditis hermaphrodita*) pests.

- Entomopathogenic viruses (*e.g. Cydia pomonella* granulovirus).

- Weeds and rodents have also been controlled with microbial agents.

Various naturally occurring materials, including fungal and plant extracts, have been described as biopesticides. Products in this category include:

- Insect pheromones and other semiochemicals.

- Fermentation products such as Spinosad (a macro-cyclic lactone).

- Chitosan: a plant in the presence of this product will naturally induce systemic resistance (ISR) to allow the plant to defend itself against disease, pathogens and pests.

- Biopesticides may include natural plant-derived products, which include alkaloids, terpenoids, phenolics and other secondary chemicals. Certain vegetable oils such as canola oil are known to have pesticidal properties. Products based on extracts of plants such as garlic have now been registered in the EU and elsewhere.

Applications

Biopesticides are biological or biologically-derived agents, that are usually applied in a manner similar to chemical pesticides, but achieve pest management in an environmentally friendly way. With all pest management products, but especially microbial agents, effective control requires appropriate formulation and application.

Biopesticides for use against crop diseases have already established themselves on a variety of crops. For example, biopesticides already play an important role in controlling downy mildew diseases. Their benefits include: a 0-Day Pre-Harvest Interval, the ability to use under moderate to severe disease pressure, and the ability to use as a tank mix or in a rotational program with other registered fungicides. Because some market studies estimate that as much as 20% of global fungicide sales are directed at downy mildew diseases, the integration of biofungicides into grape production has substantial benefits in terms of extending the useful life of other fungicides, especially those in the reduced-risk category.

A major growth area for biopesticides is in the area of seed treatments and soil amendments. Fungicidal and biofungicidal seed treatments are used to control soil borne fungal pathogens that cause seed rots, damping-off, root rot and seedling blights. They can also be used to control internal seed–borne fungal pathogens as well as fungal pathogens that are on the surface of the seed. Many biofungicidal products also show capacities to stimulate plant host defence and other physiological processes that can make treated crops more resistant to a variety of biotic and abiotic stresses.

Disadvantages

- High specificity: Which may require an exact identification of the pest/pathogen and the use of multiple products to be used; although this can also be an advantage in that the biopesticide is less likely to harm species other than the target.

- Often slow speed of action (thus making them unsuitable if a pest outbreak is an immediate threat to a crop).

- Often variable efficacy due to the influences of various biotic and abiotic factors (since some biopesticides are living organisms, which bring about pest/pathogen control by multiplying within or nearby the target pest/pathogen).

- Living organisms evolve and increase their resistance to biological, chemical, physical or any other form of control. If the target population is not exterminated or rendered incapable of reproduction, the surviving population can acquire a tolerance of whatever pressures are brought to bear, resulting in an evolutionary arms race.

- Unintended consequences: Studies have found broad spectrum biopesticides have lethal and nonlethal risks for non-target native pollinators such as *Melipona quadrifasciata* in Brazil.

HERBICIDES

Herbicides also commonly known as weedkillers, are substances used to control un-wanted plants. Selective herbicides control specific weed species, while leaving the desired crop relatively unharmed, while non-selective herbicides (sometimes called total weedkillers in commercial products) can be used to clear waste ground, industrial and construction sites, railways and railway embankments as they kill all plant material with which they come into contact. Apart from selective/non-selective, other important distinctions include *persistence* (also known as *residual action*: how long the product stays in place and remains active), *means of uptake* (whether it is absorbed by above-ground foliage only, through the roots, or by other means), and *mechanism of action* (how it works). Historically, products such as common salt and other metal salts were used as herbicides, however these have gradually fallen out of favor and in some countries a number of these are banned due to their persistence in soil, and toxicity and groundwater contamination concerns. Herbicides have also been used in warfare and conflict.

Modern herbicides are often synthetic mimics of natural plant hormones which interfere with growth of the target plants. The term organic herbicide has come to mean herbicides intended for organic farming. Some plants also produce their own natural herbicides, such as the genus Juglans (walnuts), or the tree of heaven; such action of natural herbicides, and other related chemical interactions, is called allelopathy. Due to herbicide resistance - a major concern in agriculture - a number of products combine herbicides with different means of action. Integrated pest management may use herbicides alongside other pest control methods.

In the US in 2007, about 83% of all herbicide usage, determined by weight applied, was in agriculture. In 2007, world pesticide expenditures totaled about $39.4 billion; herbicides were about 40% of those sales and constituted the biggest portion, followed by insecticides, fungicides, and other types. Herbicide is also used in forestry, where certain formulations have been found to suppress hardwood varieties in favour of conifers after a clearcut, as well as pasture systems, and management of areas set aside as wildlife habitat.

Mechanism of Action

Herbicides are often classified according to their site of action, because as a general rule, herbicides within the same site of action class will produce similar symptoms on susceptible plants. Classification based on site of action of herbicide is comparatively better as herbicide resistance management can be handled more properly and effectively. Classification by mechanism of action (MOA) indicates the first enzyme, protein, or biochemical step affected in the plant following application:

- ACCase inhibitors: Acetyl coenzyme A carboxylase (ACCase) is part of the first step of lipid synthesis. Thus, ACCase inhibitors affect cell membrane production in the meristems of the grass plant. The ACCases of grasses are sensitive to these herbicides, whereas the ACCases of dicot plants are not.

- ALS inhibitors: the acetolactate synthase (ALS) enzyme (also known as acetohydroxyacid synthase, or AHAS) is the first step in the synthesis of the branched-chain amino acids (valine, leucine, and isoleucine). These herbicides slowly starve affected plants of these amino acids, which eventually leads to inhibition of DNA synthesis. They affect grasses and dicots alike. The ALS inhibitor family includes various sulfonylureas (SUs) (such as Flazasulfuron and Metsulfuron-methyl), imidazolinones (IMIs), triazolopyrimidines (TPs), pyrimidinyl oxybenzoates (POBs), and sulfonylamino carbonyl triazolinones (SCTs). The ALS biological pathway exists only in plants and not animals, thus making the ALS-inhibitors among the safest herbicides.

- EPSPS inhibitors: Enolpyruvylshikimate 3-phosphate synthase enzyme (EPSPS) is used in the synthesis of the amino acids tryptophan, phenylalanine and tyrosine. They affect grasses and dicots alike. Glyphosate (Roundup) is a systemic EPSPS inhibitor inactivated by soil contact.

- The discovery of synthetic auxins inaugurated the era of organic herbicides. They were discovered in the 1940s after long study of the plant growth regulator auxin. Synthetic auxins mimic in some way this plant hormone. They have several points of action on the cell membrane, and are effective in the control of dicot plants. 2,4-D and 2,4,5-T are synthetic auxin herbicides.

- Photosystem II inhibitors reduce electron flow from water to $NADP^+$ at the photochemical step in photosynthesis. They bind to the Qb site on the D1 protein, and prevent quinone from binding to this site. Therefore, this group of compounds causes electrons to accumulate on chlorophyll molecules. As a consequence, oxidation reactions in excess of those normally tolerated by the cell occur, and the plant dies. The triazine herbicides (including atrazine) and urea derivatives (diuron) are photosystem II inhibitors.

- Photosystem I inhibitors steal electrons from the normal pathway through FeS to Fdx to $NADP^+$ leading to direct discharge of electrons on oxygen. As a result, reactive oxygen species are produced and oxidation reactions in excess of those normally tolerated by the cell occur, leading to plant death. Bipyridinium herbicides (such as diquat and paraquat) inhibit the FeS to Fdx step of that chain, while diphenyl ether herbicides (such as nitrofen, nitrofluorfen, and acifluorfen) inhibit the Fdx to $NADP^+$ step.

- HPPD inhibitors inhibit 4-Hydroxyphenylpyruvate dioxygenase, which are involved in tyrosine breakdown. Tyrosine breakdown products are used by plants to make carotenoids, which protect chlorophyll in plants from being destroyed by sunlight. If this happens, the plants turn white due to complete loss of chlorophyll, and the plants die. Mesotrione and sulcotrione are herbicides in this class; a drug, nitisinone, was discovered in the course of developing this class of herbicides.

Herbicide Group (Labeling)

One of the most important methods for preventing, delaying, or managing resistance is to reduce the reliance on a single herbicide mode of action. To do this, farmers must know the mode of action for the herbicides they intend to use, but the relatively complex nature of plant biochemistry makes this difficult to determine. Attempts were made to simplify the understanding of herbicide mode of action by developing a classification system that grouped herbicides by mode of action. Eventually the Herbicide Resistance Action Committee (HRAC) and the Weed Science Society of America (WSSA) developed a classification system. The WSSA and HRAC systems differ in the group designation. Groups in the WSSA and the HRAC systems are designated by numbers and letters, respectively. The goal for adding the "Group" classification and mode of action to the herbicide product label is to provide a simple and practical approach to deliver the information to users. This information will make it easier to develop educational material that is consistent and effective. It should increase user's awareness of herbicide mode of action and provide more accurate recommendations for resistance management. Another goal is to make it easier for users to keep records on which herbicide mode of actions are being used on a particular field from year to year.

Chemical Family

Detailed investigations on chemical structure of the active ingredients of the registered herbicides showed that some moieties (moiety is a part of a molecule that may include either whole functional groups or parts of functional groups as substructures; a functional group has similar chemical properties whenever it occurs in different compounds) have the same mechanisms of action. According to Forouzesh *et al.* 2015, these moieties have been assigned to the names of chemical families and active ingredients are then classified within the chemical families accordingly. Knowing about herbicide chemical family grouping could serve as a short-term strategy for managing resistance to site of action.

Use and Application

Herbicides being sprayed from the spray arms of a tractor.

Most herbicides are applied as water-based sprays using ground equipment. Ground equipment varies in design, but large areas can be sprayed using self-propelled sprayers equipped with long booms, of 60 to 120 feet (18 to 37 m) with spray nozzles spaced every 20–30 inches (510–760 mm) apart. Towed, handheld, and even horse-drawn sprayers are also used. On large areas, herbicides may also at times be applied aerially using helicopters or airplanes, or through irrigation systems (known as chemigation).

A further method of herbicide application developed around 2010, involves ridding the soil of its active weed seed bank rather than just killing the weed. This can successfully treat annual plants but not perennials. Researchers at the Agricultural Research Service found that the application of herbicides to fields late in the weeds' growing season greatly reduces their seed production, and therefore fewer weeds will return the following season. Because most weeds are annuals, their seeds will only survive in soil for a year or two, so this method will be able to destroy such weeds after a few years of herbicide application.

Weed-wiping may also be used, where a wick wetted with herbicide is suspended from a boom and dragged or rolled across the tops of the taller weed plants. This allows treatment of taller grassland weeds by direct contact without affecting related but desirable shorter plants in the grassland sward beneath. The method has the benefit of avoiding spray drift. In Wales, a scheme offering free weed-wiper hire was launched in 2015 in an effort to reduce the levels of MCPA in water courses.

Misuse and Misapplication

Herbicide volatilisation or spray drift may result in herbicide affecting neighboring fields or plants, particularly in windy conditions. Sometimes, the wrong field or plants may be sprayed due to error.

Use Politically, Militarily and in Conflict

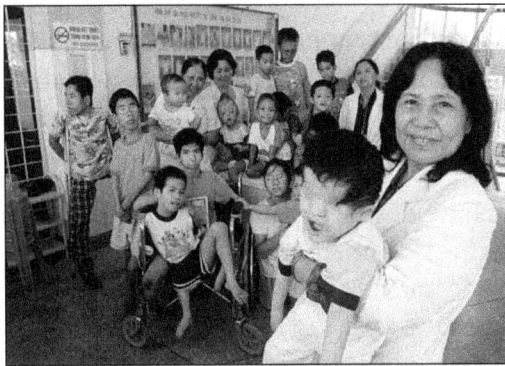

Handicapped children in Vietnam, most of them victims of Agent Orange.

Although herbicidal warfare use chemical substances, its main purpose is to disrupt agricultural food production and/or to destroy plants which provide cover or concealment to the enemy.

The use of herbicides as a chemical weapon by the U.S. military during the Vietnam War has left tangible, long-term impacts upon the Vietnamese people that live in Vietnam. For instance, it led to 3 million Vietnamese people suffering health problems, one million birth defects caused directly by exposure to Agent Orange, and 24% of the area of Vietnam being defoliated.

Health and Environmental Effects

Herbicides have widely variable toxicity in addition to acute toxicity arising from ingestion of a significant quantity rapidly, and chronic toxicity arising from environmental and occupational exposure over long periods. Much public suspicion of herbicides revolves around a confusion between valid statements of *acute* toxicity as opposed to equally valid statements of lack of *chronic* toxicity at the recommended levels of usage. For instance, while glyphosate formulations with tallowamine *adjuvants* are acutely toxic, their use was found to be uncorrelated with any health issues like cancer in a massive US Department of Health study on 90,000 members of farmer families for over a period of 23 years. That is, the study shows lack of chronic toxicity, but cannot question the herbicide's acute toxicity.

Some herbicides cause a range of health effects ranging from skin rashes to death. The pathway of attack can arise from intentional or unintentional direct consumption, improper application resulting in the herbicide coming into direct contact with people or wildlife, inhalation of aerial sprays, or food consumption prior to the labelled preharvest interval. Under some conditions, certain herbicides can be transported via leaching or surface runoff to contaminate groundwater or distant surface water sources. Generally, the conditions that promote herbicide transport include intense storm events (particularly shortly after application) and soils with limited capacity to adsorb or retain the herbicides. Herbicide properties that increase likelihood of transport include persistence (resistance to degradation) and high water solubility.

Phenoxy herbicides are often contaminated with dioxins such as TCDD; research has suggested such contamination results in a small rise in cancer risk after occupational exposure to these herbicides. Triazine exposure has been implicated in a likely relationship to increased risk of breast cancer, although a causal relationship remains unclear.

Herbicide manufacturers have at times made false or misleading claims about the safety of their products. Chemical manufacturer Monsanto Company agreed to change its advertising after pressure from New York attorney general Dennis Vacco; Vacco complained about misleading claims that its spray-on glyphosate-based herbicides, including Roundup, were safer than table salt and "practically non-toxic" to mammals, birds, and fish (though proof that this was ever said is hard to find). Roundup is toxic and has resulted in death after being ingested in quantities ranging from 85 to 200 ml, although it has also been ingested in quantities as large as 500 ml with only mild or moderate symptoms. The manufacturer of Tordon 101 (Dow AgroSciences, owned by the Dow

Chemical Company) has claimed Tordon 101 has no effects on animals and insects, in spite of evidence of strong carcinogenic activity of the active ingredient Picloram in studies on rats.

The risk of Parkinson's disease has been shown to increase with occupational exposure to herbicides and pesticides. The herbicide paraquat is suspected to be one such factor.

All commercially sold, organic and nonorganic herbicides must be extensively tested prior to approval for sale and labeling by the Environmental Protection Agency. However, because of the large number of herbicides in use, concern regarding health effects is significant. In addition to health effects caused by herbicides themselves, commercial herbicide mixtures often contain other chemicals, including inactive ingredients, which have negative impacts on human health.

Ecological Effects

Commercial herbicide use generally has negative impacts on bird populations, although the impacts are highly variable and often require field studies to predict accurately. Laboratory studies have at times overestimated negative impacts on birds due to toxicity, predicting serious problems that were not observed in the field. Most observed effects are due not to toxicity, but to habitat changes and the decreases in abundance of species on which birds rely for food or shelter. Herbicide use in silviculture, used to favor certain types of growth following clearcutting, can cause significant drops in bird populations. Even when herbicides which have low toxicity to birds are used, they decrease the abundance of many types of vegetation on which the birds rely. Herbicide use in agriculture in Britain has been linked to a decline in seed-eating bird species which rely on the weeds killed by the herbicides. Heavy use of herbicides in neotropical agricultural areas has been one of many factors implicated in limiting the usefulness of such agricultural land for wintering migratory birds.

Frog populations may be affected negatively by the use of herbicides as well. While some studies have shown that atrazine may be a teratogen, causing demasculinization in male frogs, the U.S. Environmental Protection Agency (EPA) and its independent Scientific Advisory Panel (SAP) examined all available studies on this topic and concluded that "atrazine does not adversely affect amphibian gonadal development based on a review of laboratory and field studies."

Scientific Uncertainty of Full Extent of Herbicide Effects

The health and environmental effects of many herbicides is unknown, and even the scientific community often disagrees on the risk. For example, a 1995 panel of 13 scientists reviewing studies on the carcinogenicity of 2,4-D had divided opinions on the likelihood 2,4-D causes cancer in humans. As of 1992, studies on phenoxy herbicides were too few to accurately assess the risk of many types of cancer from these herbicides, even though evidence was stronger that exposure to these herbicides is associated with increased risk of soft tissue sarcoma and non-Hodgkin lymphoma. Furthermore, there is some suggestion that herbicides can play a role in sex reversal of certain organisms that experience temperature-dependent sex determination, which could theoretically alter sex ratios.

Resistance

Weed resistance to herbicides has become a major concern in crop production worldwide. Resistance to herbicides is often attributed to lack of rotational programmes of herbicides and to continuous applications of herbicides with the same sites of action. Thus, a true understanding of the sites of action of herbicides is essential for strategic planning of herbicide-based weed control.

Plants have developed resistance to atrazine and to ALS-inhibitors, and more recently, to glyphosate herbicides. Marestail is one weed that has developed glyphosate resistance. Glyphosate-resistant weeds are present in the vast majority of soybean, cotton and corn farms in some U.S. states. Weeds that can resist multiple other herbicides are spreading. Few new herbicides are near commercialization, and none with a molecular mode of action for which there is no resistance. Because most herbicides could not kill all weeds, farmers rotated crops and herbicides to stop resistant weeds. During its initial years, glyphosate was not subject to resistance and allowed farmers to reduce the use of rotation.

A family of weeds that includes waterhemp (Amaranthus rudis) is the largest concern. A 2008-9 survey of 144 populations of waterhemp in 41 Missouri counties revealed glyphosate resistance in 69%. Weeds from some 500 sites throughout Iowa in 2011 and 2012 revealed glyphosate resistance in approximately 64% of waterhemp samples. The use of other killers to target "residual" weeds has become common, and may be sufficient to have stopped the spread of resistance. From 2005 through 2010 researchers discovered 13 different weed species that had developed resistance to glyphosate. But since then only two more have been discovered. Weeds resistant to multiple herbicides with completely different biological action modes are on the rise. In Missouri, 43% of samples were resistant to two different herbicides; 6% resisted three; and 0.5% resisted four. In Iowa 89% of waterhemp samples resist two or more herbicides, 25% resist three, and 10% resist five.

For southern cotton, herbicide costs has climbed from between $50 and $75 per hectare a few years ago to about $370 per hectare in 2013. Resistance is contributing to a massive shift away from growing cotton; over the past few years, the area planted with cotton has declined by 70% in Arkansas and by 60% in Tennessee. For soybeans in Illinois, costs have risen from about $25 to $160 per hectare.

Dow, Bayer CropScience, Syngenta and Monsanto are all developing seed varieties resistant to herbicides other than glyphosate, which will make it easier for farmers to use alternative weed killers. Even though weeds have already evolved some resistance to those herbicides, Powles says the new seed-and-herbicide combos should work well if used with proper rotation.

Biochemistry of Resistance

Resistance to herbicides can be based on one of the following biochemical mechanisms:

- Target-site resistance: This is due to a reduced (or even lost) ability of the herbicide to bind to its target protein. The effect usually relates to an enzyme with a crucial function in a metabolic pathway, or to a component of an electron-transport system. Target-site resistance may also be caused by an overexpression of the target enzyme (via gene amplification or changes in a gene promoter).

- Non-target-site resistance: This is caused by mechanisms that reduce the amount of herbicidal active compound reaching the target site. One important mechanism is an enhanced metabolic detoxification of the herbicide in the weed, which leads to insufficient amounts of the active substance reaching the target site. A reduced uptake and translocation, or sequestration of the herbicide, may also result in an insufficient herbicide transport to the target site.

- Cross-resistance: In this case, a single resistance mechanism causes resistance to several herbicides. The term target-site cross-resistance is used when the herbicides bind to the same target site, whereas non-target-site cross-resistance is due to a single non-target-site mechanism (e.g., enhanced metabolic detoxification) that entails resistance across herbicides with different sites of action.

- Multiple resistance: In this situation, two or more resistance mechanisms are present within individual plants, or within a plant population.

Resistance Management

Worldwide experience has been that farmers tend to do little to prevent herbicide resistance developing, and only take action when it is a problem on their own farm or neighbor's. Careful observation is important so that any reduction in herbicide efficacy can be detected. This may indicate evolving resistance. It is vital that resistance is detected at an early stage as if it becomes an acute, whole-farm problem, options are more limited and greater expense is almost inevitable. Table lists factors which enable the risk of resistance to be assessed. An essential pre-requisite for confirmation of resistance is a good diagnostic test. Ideally this should be rapid, accurate, cheap and accessible. Many diagnostic tests have been developed, including glasshouse pot assays, petri dish assays and chlorophyll fluorescence. A key component of such tests is that the response of the suspect population to a herbicide can be compared with that of known susceptible and resistant standards under controlled conditions. Most cases of herbicide resistance are a consequence of the repeated use of herbicides, often in association with crop monoculture and reduced cultivation practices. It is necessary, therefore, to modify these practices in order to prevent or delay the onset of resistance or to control existing resistant populations. A key objective should be the reduction in selection pressure. An integrated weed management (IWM) approach is required, in which as many tactics as possible are used to combat weeds. In this way, less reliance is placed on herbicides and so selection pressure should be reduced.

Optimising herbicide input to the economic threshold level should avoid the unnecessary use of herbicides and reduce selection pressure. Herbicides should be used to their greatest potential by ensuring that the timing, dose, application method, soil and climatic conditions are optimal for good activity. In the UK, partially resistant grass weeds such as *Alopecurus myosuroides* (blackgrass) and *Avena* spp. (wild oat) can often be controlled adequately when herbicides are applied at the 2-3 leaf stage, whereas later applications at the 2-3 tiller stage can fail badly. Patch spraying, or applying herbicide to only the badly infested areas of fields, is another means of reducing total herbicide use.

Table: Agronomic factors influencing the risk of herbicide resistance development.

Factor	Low risk	High risk
Cropping system	Good rotation	Crop monoculture
Cultivation system	Annual ploughing	Continuous minimum tillage
Weed control	Cultural only	Herbicide only
Herbicide use	Many modes of action	Single modes of action
Control in previous years	Excellent	Poor
Weed infestation	Low	High
Resistance in vicinity	Unknown	Common

Approaches to Treating Resistant Weeds

Alternative Herbicides

When resistance is first suspected or confirmed, the efficacy of alternatives is likely to be the first consideration. The use of alternative herbicides which remain effective on resistant populations can be a successful strategy, at least in the short term. The effectiveness of alternative herbicides will be highly dependent on the extent of cross-resistance. If there is resistance to a single group of herbicides, then the use of herbicides from other groups may provide a simple and effective solution, at least in the short term. For example, many triazine-resistant weeds have been readily controlled by the use of alternative herbicides such as dicamba or glyphosate. If resistance extends to more than one herbicide group, then choices are more limited. It should not be assumed that resistance will automatically extend to all herbicides with the same mode of action, although it is wise to assume this until proved otherwise. In many weeds the degree of cross-resistance between the five groups of ALS inhibitors varies considerably. Much will depend on the resistance mechanisms present, and it should not be assumed that these will necessarily be the same in different populations of the same species. These differences are due, at least in part, to the existence of different mutations conferring target site resistance. Consequently, selection for different mutations may result in different patterns of cross-resistance. Enhanced metabolism can affect even closely related herbicides to differing degrees. For example, populations of *Alopecurus myosuroides* (blackgrass) with an enhanced metabolism mechanism show resistance to pendimethalin but not to trifluralin, despite both being dinitroanilines. This is due to differences in the vulnerability of these two herbicides to oxidative metabolism. Consequently, care is needed when trying to predict the efficacy of alternative herbicides.

Mixtures and Sequences

The use of two or more herbicides which have differing modes of action can reduce the selection for resistant genotypes. Ideally, each component in a mixture should:

- Be active at different target sites.

- Have a high level of efficacy.

- Be detoxified by different biochemical pathways.

- Have similar persistence in the soil (if it is a residual herbicide).

- Exert negative cross-resistance.

- Synergise the activity of the other component.

No mixture is likely to have all these attributes, but the first two listed are the most important. There is a risk that mixtures will select for resistance to both components in the longer term. One practical advantage of sequences of two herbicides compared with mixtures is that a better appraisal of the efficacy of each herbicide component is possible, provided that sufficient time elapses between each application. A disadvantage with sequences is that two separate applications have to be made and it is possible that the later application will be less effective on weeds surviving the first application. If these are resistant, then the second herbicide in the sequence may increase selection for resistant individuals by killing the susceptible plants which were damaged but not killed by the first application, but allowing the larger, less affected, resistant plants to survive. This has been cited as one reason why ALS-resistant *Stellaria media* has evolved in Scotland recently, despite the regular use of a sequence incorporating mecoprop, a herbicide with a different mode of action.

Herbicide Rotations

Rotation of herbicides from different chemical groups in successive years should reduce selection for resistance. This is a key element in most resistance prevention programmes. The value of this approach depends on the extent of cross-resistance, and whether multiple resistance occurs owing to the presence of several different resistance mechanisms. A practical problem can be the lack of awareness by farmers of the different groups of herbicides that exist. In Australia a scheme has been introduced in which identifying letters are included on the product label as a means of enabling farmers to distinguish products with different modes of action.

Farming Practices and Resistance

Herbicide resistance became a critical problem in Australian agriculture, after many Australian sheep farmers began to exclusively grow wheat in their pastures in the 1970s. Introduced varieties of ryegrass, while good for grazing sheep, compete intensely with wheat. Ryegrasses produce so many seeds that, if left unchecked, they can completely choke a field. Herbicides provided excellent control, while reducing soil disrupting because of less need to plough. Within little more than a decade, ryegrass and other weeds began to develop resistance. In response Australian farmers changed methods. By 1983, patches of ryegrass had become immune to Hoegrass, a family of herbicides that inhibit an enzyme called acetyl coenzyme A carboxylase.

Ryegrass populations were large, and had substantial genetic diversity, because farmers had planted many varieties. Ryegrass is cross-pollinated by wind, so genes shuffle frequently. To control its distribution farmers sprayed inexpensive Hoegrass, creating selection pressure. In addition, farmers sometimes diluted the herbicide in order to save money, which allowed some plants to survive application. When resistance appeared farmers turned to a group of herbicides that block acetolactate synthase. Once again, ryegrass in Australia evolved a kind of "cross-resistance" that allowed it to rapidly break down a variety of herbicides. Four classes of herbicides become ineffective within a few years. In 2013 only two herbicide classes, called Photosystem II and long-chain fatty acid inhibitors, were effective against ryegrass.

Common Herbicides

Synthetic Herbicides

- 2,4-D is a broadleaf herbicide in the phenoxy group used in turf and no-till field crop production. Now, it is mainly used in a blend with other herbicides to allow lower rates of herbicides to be used; it is the most widely used herbicide in the world, and third most commonly used in the United States. It is an example of synthetic auxin (plant hormone).

- Aminopyralid is a broadleaf herbicide in the pyridine group, used to control weeds on grassland, such as docks, thistles and nettles. It is notorious for its ability to persist in compost.

- Atrazine, a triazine herbicide, is used in corn and sorghum for control of broadleaf weeds and grasses. Still used because of its low cost and because it works well on a broad spectrum of weeds common in the US corn belt, atrazine is commonly used with other herbicides to reduce the overall rate of atrazine and to lower the potential for groundwater contamination; it is a photosystem II inhibitor.

- Clopyralid is a broadleaf herbicide in the pyridine group, used mainly in turf, rangeland, and for control of noxious thistles. Notorious for its ability to persist in compost, it is another example of synthetic auxin.

- Dicamba, a postemergent broadleaf herbicide with some soil activity, is used on turf and field corn. It is another example of a synthetic auxin.

- Glufosinate ammonium, a broad-spectrum contact herbicide, is used to control weeds after the crop emerges or for total vegetation control on land not used for cultivation.

- Fluazifop (Fuselade Forte), a post emergence, foliar absorbed, translocated grass-selective herbicide with little residual action. It is used on a very wide range of broad leaved crops for control of annual and perennial grasses.

- Fluroxypyr, a systemic, selective herbicide, is used for the control of broad-leaved weeds in small grain cereals, maize, pastures, rangeland and turf. It is a synthetic auxin. In cereal growing, fluroxypyr's key importance is control of cleavers, *Galium aparine*. Other key broadleaf weeds are also controlled.

- Glyphosate, a systemic nonselective herbicide, is used in no-till burndown and for weed control in crops genetically modified to resist its effects. It is an example of an EPSPs inhibitor.

- Imazapyr a nonselective herbicide, is used for the control of a broad range of weeds, including terrestrial annual and perennial grasses and broadleaf herbs, woody species, and riparian and emergent aquatic species.

- Imazapic, a selective herbicide for both the pre- and postemergent control of some annual and perennial grasses and some broadleaf weeds, kills plants by inhibiting the production of branched chain amino acids (valine, leucine, and isoleucine), which are necessary for protein synthesis and cell growth.

- Imazamox, an imidazolinone manufactured by BASF for postemergence application that is an acetolactate synthase (ALS) inhibitor. Sold under trade names Raptor, Beyond, and Clearcast.

- Linuron is a nonselective herbicide used in the control of grasses and broadleaf weeds. It works by inhibiting photosynthesis.

- MCPA (2-methyl-4-chlorophenoxyacetic acid) is a phenoxy herbicide selective for broadleaf plants and widely used in cereals and pasture.

- Metolachlor is a pre-emergent herbicide widely used for control of annual grasses in corn and sorghum; it has displaced some of the atrazine in these uses.

- Paraquat is a nonselective contact herbicide used for no-till burndown and in aerial destruction of marijuana and coca plantings. It is more acutely toxic to people than any other herbicide in widespread commercial use.

- Pendimethalin, a pre-emergent herbicide, is widely used to control annual grasses and some broad-leaf weeds in a wide range of crops, including corn, soybeans, wheat, cotton, many tree and vine crops, and many turfgrass species.

- Picloram, a pyridine herbicide, mainly is used to control unwanted trees in pastures and edges of fields. It is another synthetic auxin.

- Sodium chlorate *(disused/banned in some countries)*, a nonselective herbicide, is considered phytotoxic to all green plant parts. It can also kill through root absorption.

- Triclopyr, a systemic, foliar herbicide in the pyridine group, is used to control broadleaf weeds while leaving grasses and conifers unaffected.

- Several sulfonylureas, including Flazasulfuron and Metsulfuron-methyl, which act as ALS inhibitors and in some cases are taken up from the soil via the roots.

Organic Herbicides

Recently, the term "organic" has come to imply products used in organic farming. Under this definition, an organic herbicide is one that can be used in a farming enterprise that has been classified as organic. Depending on the application, they may be less effective than synthetic herbicides and are generally used along with cultural and mechanical weed control practices.

Homemade organic herbicides include:

- Corn gluten meal (CGM) is a natural pre-emergence weed control used in turfgrass, which reduces germination of many broadleaf and grass weeds.

- Vinegar is effective for 5–20% solutions of acetic acid, with higher concentrations most effective, but it mainly destroys surface growth, so respraying to treat regrowth is needed. Resistant plants generally succumb when weakened by respraying.

- Steam has been applied commercially, but is now considered uneconomical and inadequate. It controls surface growth but not underground growth and so respraying to treat regrowth of perennials is needed.

- Flame is considered more effective than steam, but suffers from the same difficulties.

- D-limonene (citrus oil) is a natural degreasing agent that strips the waxy skin or cuticle from weeds, causing dehydration and ultimately death.

- Saltwater or salt applied in appropriate strengths to the rootzone will kill most plants.

BIOHERBICIDE

Bioherbicides consist of phytotoxins, pathogens, and other microbes used as biological weed control. Bioherbicides may be compounds and secondary metabolites derived from microbes such as fungi, bacteria or protozoa; or phytotoxic plant residues, extracts or single compounds derived from other plant species.

Available Bioherbicides

While 13 different products have been launched, currently only 9 bioherbicides are available for sale/purchase in market globally. Below is the list of available bioherbicides:

1. Devine

2. Collego

3. BioMal

4. Woad Warrior

5. Chontrol

6. Smoulder

7. Sarritor

8. Organo-Sol

9. Beloukha

Marketing

With increasing awareness of the effects of the chemical herbicides and pesticides, bioherbicides can be adopted as an alternative especially for integrated weed management. The market share of bioherbicides is merely 10% of all biopesticides. On the other hand, the research spanning over two decades since 1980s has also falsified the principle that there is a coevolved natural enemy of a host weed which can manage weed through varied formulation and thus advocated for more research to culturally and genetically intensify the bioherbicidal organisms. Efficiency and efficacy of bioherbicides is impeded by changing weather and temperature and this can further obstruct the application and integration of bioherbicides. A study shows that by covering with jute turf, which retains moistures and allows one third of the sunlight to pass through, can increase the efficiency of bioherbicides and also remove some of the hindrances from the commercialization and marketing of bio-herbicides.

Production

The production of bioherbicides is a process of biosynthesis where different mediums ranging from soybean bran to corn steep liquor are fermented to obtain desirable re-sults. In addition to the solid-state fermentation, bioherbicides can also be produced by submerged fermentation in stirred tanks or in other environments. Despite the 'eco-friendliness', there are several obstructions that make it less practical to use bio-herbicides in fields because the lab results may not be the same as the real results.

BROMACIL

Bromacil is an organic compound with the chemical formula $C_9H_{13}BrN_2O_2$, commercially available as a herbicide. Bromacil was first registered as a pesticide in the U.S. in 1961. It is used for brush control and non-cropland areas. It works by interfering with photosynthesis by entering the plant through the root zone and moving throughout the plant. Bromacil is one of a group of compounds called substituted uracils. These materials are broad spectrum herbicides used for nonselective weed and brush control on non-croplands, as well as for selective weed control on a limited number of crops, such as citrus fruit and pineapple. Bromacil is also found to be excellent at controlling perennial grasses.

Safety

There are quite a few safety precautions that should be taken when dealing with Bromacil. Dry formulations containing bromacil must bear the word "Caution" and liquid formu-las must signal "Warning." Care should be exercised when spraying Bromacil on plants because it will also stop the photosynthesis of the adjacent non-target plants, therefore killing them. Bromacil should never be used in residential or recreation areas for risk of exposure. Bromacil is slightly toxic if individuals accidentally eat or touch residues and practically nontoxic if inhaled. Bromacil is a mild eye irritant and a very slight skin irri-tant. It is not a skin sensitizer. In studies using laboratory animals, bromacil is slightly toxic by the oral, dermal, and inhalation routes and has been placed in Toxicity Category IV (the lowest of four categories) for these effects. This herbicide should be stored in a cool, dry place and after any handling a thorough hand-washing is advised.

In regards to occupational exposure, the National Institute for Occupational Safety and Health has recommended workers handing bromacil not exceed an exposure of 1 ppm (10 mg/m³) over an eight-hour time-weighted average.

Applications

Bromacil is applied mainly by sprayers including boom, hand-held, knapsack, compressed air, tank-type, and power sprayers. Bromacil is also applied using aerosol, shaker, or sprinkler cans. Solid forms of bromacil are spread using granule applicators and spreaders. Application using aircraft is allowed only for Special Local Need registrations to control vegetation.

BENTAZON

Bentazon (Bentazone, Basagran, Herbatox, Leader, Laddock) is a chemical manufactured by BASF Chemicals for use in herbicides. It is categorized under the thiadiazine group of chemicals. Sodium bentazon is available commercially and appears slightly brown in colour.

Usage

Bentazon is a selective herbicide as it only damages plants unable to metabolize the chemical. It is considered safe for use on alfalfa, beans (with the exception of garbanzo beans), maize, peanuts, peas (with the exception of blackeyed peas), pepper, peppermint, rice, sorghum, soybeans and spearmint; as well as lawns and turf. Bentazon is usually applied aerially or through contact spraying on food crops to control the spread of weeds occurring amongst food crops. Herbicides containing bentazon should be kept away from high heat as it will release toxic sulfur and nitrogen fumes.

Bentazon is currently registered for use in the United States in accordance with requirements set forth by the United States Environmental Protection Agency. However as of September 2010, the herbicides Basagran M60, Basagran DF, Basagran AG, Prompt 5L and Laddock 5L are currently under review for pending requests for voluntary registration cancellation.

Water and Ground Contamination

In general, bentazon is quickly metabolized and degraded by both plants and animals. However, soil leaching and runoff is a major concern in terms of water contamination. In 1995 the Environmental Protection Agency (EPA) stated that levels of bentazon in both ground water and surface water "exceed levels of concern". Despite the establishment of a 20 parts per billion Health Advisory Level there is no requirement to measure for bentazon in water supplies as the Safe Drinking Water Act does not regulate bentazon. The United States EPA found bentazon in 64 out of 200 wells in California - the highest number of detections in their 1995 study. This prompted the State of California to review existing toxicology studies and establish a "Public Health Goal" that limits bentazon in drinking water to 200 parts per billion.

The EPA requires ground water and environmental hazard advisory labels on all commercially available herbicides containing bentazon. Both statements warn against application and/or disposal of bentazon directly into water, or in areas where soil leaching is common.

Food Contamination

A number of limits have been placed on bentazon to reduce the possibility of toxic effects on humans. Tolerance levels vary depending on the use of the food/animal product. The following tolerance levels for bentazon have been established in the United States:

- 0.02 ppm for milk.

- 0.05 ppm (parts per million) for meat and animal byproducts (poultry, eggs, cattle, hogs, sheep and goats).

- 0.05 ppm for dried beans (excluding soybeans), corn (fresh and grain), bohemian chili peppers, peanuts, rice, soybeans, and sorghum used for fodder and grain.

- 0.5 ppm for succulent beans and peas.

- 0.3 ppm for peanut hulls.

- 1 ppm for mint and dried peas.

- 3 ppm for rice (straw), corn for fodder and forage, and peanuts used in hay and forage.

- 8 ppm for pea vine hays (dried), and soybeans used for foraging or hay.

It is recommended that food and feed supplies be stored away from herbicides containing bentazon. Aerial spraying should be conducted in a manner that prevents spray drift towards water sources and food crops susceptible to bentazon.

Toxicity to Nonhuman Species

A 1994 study concluded that bentazon is non-toxic to honeybees, and is not harmful to beetles. Studies have found that bentazon is toxic to rainbow trout and bluegill sunfish at 190 ppm and 616 ppm, respectively. Bentazon is considered toxic to birds as it affects their reproductive capacities.

Among mammals, bentazon is found to be moderately toxic when ingested or absorbed through the skin. Lethal doses (LD50, the dose required to kill half the population being studied) for bentazon have been established for:

- Cats: 500 mg/kg,

- Rats: 1100 mg/kg to 2063 mg/kg,

- Mice: 400 mg/kg,

- Rabbits: 750 mg/kg.

Dogs in a study being fed 13.1 mg of bentazon a day developed diarrhoea, anemia and dehydration. In another study using dogs, prostate inflammation was also observed along with previously noted health effects. In experiments conducted on hamsters, mice and rats, bentazon was not found to cause gene mutations to damage to DNA and chromosomes.

Toxicity to Humans

Bentazon has been classified by the EPA as a "Group E" chemical, because it is believed to be non-carcinogenic to humans (as based on testing conducted on animals). However, there are no studies or experiments that can determine toxic and/or carcinogenic effects of bentazon on humans. Workers applying the herbicide would be most exposed to bentazon, and so have been advised to wear protective clothing (goggles, gloves and aprons) at all times when handling the chemical. Bentazon causes allergy-like symptoms as it irritates the eyes, skin and respiratory tract. Ingesting bentazon causes nausea, diarrhoea, trembling, vomiting and difficulty breathing. Workers handling bentazon must wash their hands before eating, drinking, smoking, and using the bathroom to minimize contact with skin. The effects of bentazon ingestion has been observed in humans who chose the herbicide to commit suicide. Ingestion of bentazon was observed to cause fevers, renal failure (kidney failure), accelerated heart rate (tachycardia), shortness of breath (dyspnea) and hyperthermia. Ingestion of 88 grams of bentazon caused death in an adult.

FUNGICIDES

Fungicides are biocidal chemical compounds or biological organisms used to kill parasitic fungi or their spores. A fungistatic inhibits their growth. Fungi can cause serious damage in agriculture, resulting in critical losses of yield, quality, and profit. Fungicides are used both in agriculture and to fight fungal infections in animals. Chemicals used to control oomycetes, which are not fungi, are also referred to as fungicides, as oomycetes use the same mechanisms as fungi to infect plants.

Fungicides can either be contact, translaminar or systemic. Contact fungicides are not taken up into the plant tissue and protect only the plant where the spray is deposited. Translaminar fungicides redistribute the fungicide from the upper, sprayed leaf surface to the lower, unsprayed surface. Systemic fungicides are taken up and redistributed through the xylem vessels. Few fungicides move to all parts of a plant. Some are locally systemic, and some move upwardly.

Most fungicides that can be bought retail are sold in a liquid form. A very common active ingredient is sulfur, present at 0.08% in weaker concentrates, and as high as 0.5% for more potent fungicides. Fungicides in powdered form are usually around 90% sulfur and are very toxic. Other active ingredients in fungicides include neem oil, rosemary oil, jojoba oil, the bacterium *Bacillus subtilis*, and the beneficial fungus *Ulocladium oudemansii*.

Fungicide residues have been found on food for human consumption, mostly from post-harvest treatments. Some fungicides are dangerous to human health, such as vinclozolin, which has now been removed from use. Ziram is also a fungicide that is toxic to humans with long-term exposure, and fatal if ingested. A number of fungicides are also used in human health care.

Natural Fungicides

Plants and other organisms have chemical defenses that give them an advantage against microorganisms such as fungi. Some of these compounds can be used as fungicides:

- Tea tree oil,

- Citronella oil,

- Jojoba oil,

- Nimbin,

- Oregano oil,

- Rosemary oil,

- Monocerin,

- Milk.

Whole live or dead organisms that are efficient at killing or inhibiting fungi can sometimes be used as fungicides:

- Bacillus subtilis,

- Ulocladium oudemansii,

- Kelp (powdered dried kelp is fed to cattle to help prevent fungal infection),

- *Ampelomyces quisqualis.*

Resistance

Pathogens respond to the use of fungicides by evolving resistance. In the field several mechanisms of resistance have been identified. The evolution of fungicide resistance can be gradual or sudden. In qualitative or discrete resistance, a mutation (normally to a single gene) produces a race of a fungus with a high degree of resistance. Such resistant varieties also tend to show stability, persisting after the fungicide has been removed from the market. For example, sugar beet leaf blotch remains resistant to azoles years after they were no longer used for control of the disease. This is because such mutations have a high selection pressure when the fungicide is used, but there is low selection pressure to remove them in the absence of the fungicide.

In instances where resistance occurs more gradually, a shift in sensitivity in the pathogen to the fungicide can be seen. Such resistance is polygenic – an accumulation of many mutations in different genes, each having a small additive effect. This type of resistance is known as quantitative or continuous resistance. In this kind of resistance, the pathogen population will revert to a sensitive state if the fungicide is no longer applied.

Little is known about how variations in fungicide treatment affect the selection pressure to evolve resistance to that fungicide. Evidence shows that the doses that provide the most control of the disease also provide the largest selection pressure to acquire resistance, and that lower doses decrease the selection pressure.

In some cases when a pathogen evolves resistance to one fungicide, it automatically obtains resistance to others – a phenomenon known as cross resistance. These additional fungicides are normally of the same chemical family or have the same mode of action, or can be detoxified by the same mechanism. Sometimes negative cross resistance occurs, where resistance to one chemical class of fungicides leads to an increase in sensitivity to a different chemical class of fungicides. This has been seen with carbendazim and diethofencarb.

There are also recorded incidences of the evolution of multiple drug resistance by pathogens – resistance to two chemically different fungicides by separate mutation events. For example, *Botrytis cinerea* is resistant to both azoles and dicarboximide fungicides.

There are several routes by which pathogens can evolve fungicide resistance. The most common mechanism appears to be alteration of the target site, in particular as a defence against single site of action fungicides. For example, Black Sigatoka, an economically important pathogen of banana, is resistant to the QoI fungicides, due to a single nucleotide change resulting in the replacement of one amino acid (glycine) by another (alanine) in the target protein of the QoI fungicides, cytochrome b. It is presumed that this disrupts the binding of the fungicide to the protein, rendering the fungicide ineffective. Upregulation of target genes can also render the fungicide ineffective. This is seen in DMI-resistant strains of *Venturia inaequalis*.

Resistance to fungicides can also be developed by efficient efflux of the fungicide out of the cell. *Septoria tritici* has developed multiple drug resistance using this mechanism. The pathogen had five ABC-type transporters with overlapping substrate specificities that together work to pump toxic chemicals out of the cell.

In addition to the mechanisms outlined above, fungi may also develop metabolic pathways that circumvent the target protein, or acquire enzymes that enable metabolism of the fungicide to a harmless substance.

Fungicide Resistance Management

The fungicide resistance action committee (FRAC) has several recommended practices to try to avoid the development of fungicide resistance, especially in at-risk fungicides including *Strobilurins* such as azoxystrobin.

Products should not be used in isolation, but rather as mixture, or alternate sprays, with another fungicide with a different mechanism of action. The likelihood of the pathogen's developing resistance is greatly decreased by the fact that any resistant isolates to one fungicide will be killed by the other; in other words, two mutations would be required rather than just one. The effectiveness of this technique can be demonstrated by Metalaxyl, a phenylamide fungicide. When used as the sole product in Ireland to control potato blight (*Phytophthora infestans*), resistance developed within one growing season. However, in countries like the UK where it was marketed only as a mixture, resistance problems developed more slowly.

Fungicides should be applied only when absolutely necessary, especially if they are in an at-risk group. Lowering the amount of fungicide in the environment lowers the selection pressure for resistance to develop.

Manufacturers' doses should always be followed. These doses are normally designed to give the right balance between controlling the disease and limiting the risk of resistance development. Higher doses increase the selection pressure for single-site mutations that confer resistance, as all strains but those that carry the mutation will be eliminated, and thus the resistant strain will propagate. Lower doses greatly increase the risk of polygenic resistance, as strains that are slightly less sensitive to the fungicide may survive.

It is better to use an integrative pest management approach to disease control rather than relying on fungicides alone. This involves the use of resistant varieties and hygienic practices, such as the removal of potato discard piles and stubble on which the pathogen can overwinter, greatly reducing the titre of the pathogen and thus the risk of fungicide resistance development.

AGRICULTURAL LIME

Agricultural lime, also called aglime, agricultural limestone, garden lime or liming, is a soil additive made from pulverized limestone or chalk. The primary active component is calcium carbonate. Additional chemicals vary depending on the mineral source and may include calcium oxide. Unlike the types of lime called quicklime (calcium oxide) and slaked lime (calcium hydroxide), powdered limestone does not require lime burning in a lime kiln; it only requires milling.

The effects of agricultural lime on soil are:

- It increases the pH of acidic soil (the lower the pH the more acidic the soil); in other words, soil acidity is reduced and alkalinity increased.

- It provides a source of calcium and magnesium for plants.

- It permits improved water penetration for acidic soils.

- It improves the uptake of major plant nutrients (nitrogen, phosphorus, and potassium) of plants growing on acid soils.

Lime may occur naturally in some soils but may require addition of sulfuric acid for its agricultural benefits to be realized. Gypsum is also used to supply calcium for plant nutrition. The concept of "corrected lime potential" to define the degree of base saturation in soils became the basis for procedures now used in soil testing laboratories to determine the "lime requirement" of soils.

Other forms of lime have common applications in agriculture and gardening, including dolomitic lime and hydrated lime. Dolomitic lime may be used as a soil input to provide similar effects as agricultural lime, while supplying magnesium in addition to calcium. In livestock farming, hydrated lime can be used as a disinfectant measure, producing a dry and alkaline environment in which bacteria do not readily multiply. In horticultural farming it can be used as an insect repellent, without causing harm to the pest or plant.

Spinner-style lime spreaders are generally used to spread agricultural lime on fields.

Agricultural lime is injected into coal burners at power plants to reduce the pollutants such as NO_2 and SO_2 from the emissions.

Determining the Need for Agricultural Lime

This is vital to maximise crop yield, animal grazing and good quality silage/hay. Soils become acidic in a number of ways. Locations that have high rainfall levels become acidic through leaching. Land used for crop and livestock purposes lose minerals over time by crop removal and becomes acidic. For example, when a 600-pound calf is removed from a pasture, 100 pounds of bone is also removed, which is 60% calcium compounds. The application of modern chemical fertilizers is a major contributor to soil acid by the process in which the plant nutrients react in the soil.

Aglime, which is high in calcium, can also be beneficial to soils where the land is used for breeding and raising foraging animals. Bone growth is key to a young animal's development and bones are composed primarily of calcium and phosphorus. Young mammals get their needed calcium through milk, which has calcium as one of its major components. Dairymen frequently apply aglime because it increases milk production.

The best way to determine if a soil is acid or deficient in calcium or magnesium is with a soil test which can be provided by a university with an agricultural education department for under $30.00, if you live in the United States. Farmers typically become interested in soil testing when they notice a decrease in crop response to applied fertilizer.

Quality

The quality of agricultural limestone is determined by the chemical makeup of the limestone and how finely the stone is ground. To aid the farmer in determining the relative value of competing agricultural liming materials, the agricultural extension services of several universities use two rating systems. Calcium Carbonate Equivalent (CCE) and the Effective Calcium Carbonate Equivalent (ECCE) give a numeric value to the effectiveness of different liming materials.

The CCE compares the chemistry of a particular quarry's stone with the neutralizing power of pure calcium carbonate. Because each molecule of magnesium carbonate is lighter than calcium carbonate, limestones containing magnesium carbonate (dolomite) can have a CCE greater than 100 percent.

Because the acids in soil are relatively weak, agricultural limestones must be ground to a small particle size to be effective. The extension service of different states rate the effectiveness of stone size particles slightly differently. They all agree, however, that the smaller the particle size the more effective the stone is at reacting in the soil. Measuring the size of particles is based on the size of a mesh that the limestone would pass through. The mesh size is the number of wires per inch. Stone retained on an 8 mesh will be about the size of BB pellets. Material passing a 60 mesh screen will have the appearance of face powder. Particles larger than 8 mesh are of little or no value, particles between 8 mesh and 60 mesh are somewhat effective and particles smaller than 60 mesh are 100 percent effective.

By combining the chemistry of a particular product (CCE) and its particle size the Effective Calcium Carbonate Equivalent (ECCE) is determined. The ECCE is percentage comparison of a particular agricultural limestone with pure calcium carbonate with all particles smaller than 60 mesh. Typically the aglime materials in commercial use will have ECCE ranging from 45 percent to 110 percent.

NEMATICIDE

Nematodes are parasitic worms that feed on living material. They can often be detrimental to plant growth and health as they attack and feed on plant roots. There are several different nematicides that are used by gardeners to control nematodes.

Although these parasitic worms that live in the soil and water are microscopic in size, they can cause major damage when they feed on plant tissues and roots. Nematicides can kill nematodes and prevent your plants from experiencing stunted growth.

One example of this is fumigant nematicides. This type of pesticide is dispersed through the spaces in soil as a gas, thus killing nematodes living in those spaces. These types of nematicides are most effective in well-drained soil.

Another type of nematicide is the non-fumigant variety. These pesticides are non-volatile, come in liquid and solid form, and can be applied to the surface or mixed into the top soil.

Most nematicides are generally only approved for commercial applications since they are highly toxic. However, there are several natural methods that can be used by gardeners to help prevent nematodes. Some natural nematicides are the use of companion plants, such as marigolds, to help control nematode populations. Additionally, soil steaming is an efficient nematicide; however, it will indiscriminately kill both harmful and beneficial soil organisms.

References

- Agrochemical, earth-and-planetary-sciences: sciencedirect.com, Retrieved 31 March, 2019

- Trapp, s.; croteau, r. (2001). "defensive biosynthesis of resin in conifers". Annual review of plant physiology and plant molecular biology. 52 (1): 689–724. Doi:10.1146/annurev.arplant.52.1.689. Pmid 11337413

- Palmer, we, bromley, pt, and brandenburg, rl. Wildlife & pesticides - peanuts. North carolina cooperative extension service. Retrieved on 14 october 2007

- Miller gt (2002). Living in the environment (12th ed.). Belmont: wadsworth/thomson learning. Isbn 9780534376970. Oclc 819417923

- Goldman lr (2007). "managing pesticide chronic health risks: u.s. policies". Journal of agromedicine. 12 (1): 67–75. Doi:10.1300/j096v12n02-08. Pmid 18032337

- American chemical society (2009). "new 'green' pesticides are first to exploit plant defenses in battle of the fungi". Eurekalert!. Aaas. Retrieved dec 1, 2018

- Robbins, paul (2007-08-27). Encyclopedia of environment and society. Robbins, paul, 1967-, sage publications. Thousand oaks. P. 862. Isbn 9781452265582. Oclc 228071686

- Nematicide, definition: maximumyield.com, Retrieved 14 July, 2019

- Zhou q, liu w, zhang y, liu kk (oct 2007). "action mechanisms of acetolactate synthase-inhibiting herbicides". Pesticide biochemistry and physiology. 89 (2): 89–96. Doi:10.1016/j.pestbp.2007.04.004

Permissions

All chapters in this book are published with permission under the Creative Commons Attribution Share Alike License or equivalent. Every chapter published in this book has been scrutinized by our experts. Their significance has been extensively debated. The topics covered herein carry significant information for a comprehensive understanding. They may even be implemented as practical applications or may be referred to as a beginning point for further studies.

We would like to thank the editorial team for lending their expertise to make the book truly unique. They have played a crucial role in the development of this book. Without their invaluable contributions this book wouldn't have been possible. They have made vital efforts to compile up to date information on the varied aspects of this subject to make this book a valuable addition to the collection of many professionals and students.

This book was conceptualized with the vision of imparting up-to-date and integrated information in this field. To ensure the same, a matchless editorial board was set up. Every individual on the board went through rigorous rounds of assessment to prove their worth. After which they invested a large part of their time researching and compiling the most relevant data for our readers.

The editorial board has been involved in producing this book since its inception. They have spent rigorous hours researching and exploring the diverse topics which have resulted in the successful publishing of this book. They have passed on their knowledge of decades through this book. To expedite this challenging task, the publisher supported the team at every step. A small team of assistant editors was also appointed to further simplify the editing procedure and attain best results for the readers.

Apart from the editorial board, the designing team has also invested a significant amount of their time in understanding the subject and creating the most relevant covers. They scrutinized every image to scout for the most suitable representation of the subject and create an appropriate cover for the book.

The publishing team has been an ardent support to the editorial, designing and production team. Their endless efforts to recruit the best for this project, has resulted in the accomplishment of this book. They are a veteran in the field of academics and their pool of knowledge is as vast as their experience in printing. Their expertise and guidance has proved useful at every step. Their uncompromising quality standards have made this book an exceptional effort. Their encouragement from time to time has been an inspiration for everyone.

The publisher and the editorial board hope that this book will prove to be a valuable piece of knowledge for students, practitioners and scholars across the globe.

Index

www.ingramcontent.com/pod-product-compliance
Lightning Source LLC
Chambersburg PA
CBHW061959190326
41458CB00009B/2911